Science Blundering
An Outsider's View

Science Blundering
An Outsider's View

by Herbert L. Nichols, Jr.

NORTH CASTLE BOOKS, INC.
Greenwich, Connecticut, U.S.A.

Designer
Joel B. Leneker

Illustrators
Albert Roth
Alice Roth Lowery

Jacket design
Robert Lowery, Jr.

Compositor
Lettick Typografic

Printing and Binding
Rose Printing Company

Copyright © 1984 by Herbert L. Nichols, Jr.

ISBN number: 0-911040-18-8

Library of Congress Number: 84-06140

Printed in the United States of America

To John,
and to Thomas
with his doubts

Dear Boys:

In our many late-night discussions, you will recall that I have urged you again and again to be distrustful. Our world has been increasingly plagued with mass-enthusiasm theories, with partly reported facts, and with warfare between prejudice and bias.

Journals of the highest rating, scientists and engineers with series of worthy accomplishments behind them, all may be (to say it politely) in error.

I have expanded notes and recollections of our talks into this small book. It is dedicated to you two, as you and your arguments have inspired so much of it.

I apologize for involving you in such a very public presentation of matters that we discussed privately, particularly as we were not always in agreement. I hope you will understand and forgive.

As you know, I have refrained from much discussion of the material in this book since it became a definite project. This was partly to keep my mind clear for simple-as-possible presentation, but mostly so that you need not feel personally involved in any misfortunes it may suffer.

We know that scientists and engineers have done a wonderful job for the human race during the last few hundred years. They have brought us knowledge and power and comforts greater than anything dreamed by our ancestors. With some exceptions, it is not their fault that the results, in behavior, life style, happiness, and future prospects, are often regrettable.

For many reasons, the relationship between scientists and non-scientists is an uneasy one. Although for-the-public scientific magazines are increasingly popular, communication is not good. Their material is largely accounts of scientific research, findings, and ideas; but much of the writing is done by non-scientists who understand the subject poorly, or scientists who do not explain it well. All too often, the material itself is unsound, representing biased views, run-with-the-crowd enthusiasms, and research and thinking that overlook major and even determining facts related to the subject.

There is a "publish or perish" situation. Many scientists are in positions where they are expected to have reports or articles published in scientific or semi-scientific magazines. Success in such publishing results in increase in prestige for both the author and his institution, and may result in his advancement. Results of not being published are correspondingly unfavorable.

Magazines must consider their own welfare. At least in the semi-scientific class, they may or must consider not only the basic value of reports, but also the extent of reader interest and acceptance.

A generation ago, articles pointing out differences between or among human races, and those citing similarities between humans and animals, were both warmly welcomed for publication. Now, the requirement is sternly opposite—racial similarities must be emphasized, human-animal resemblances minimized.

We have probably become much less accurate on the animal side, and closer to the truth on the races, but this is incidental. The facts have not changed—it is bias imposed by fashion that has.

Science itself has internal problems, sometimes arising from a seemingly desperate need to have an explanation for everything, whether it makes sense or not.

Although the title mentions only Science, the text ranges further into engineering, social movements that claim scientific justification, and even a little into industrial practice, where an absurdity invites remark.

In this book we peer together into a twilight zone, where inheritance may be an ugly word, digestive systems do not influence body weight, radiation moving at the speed of light has hung like dust in space since Creation, and weak, blind, and undirected forces are given sole credit for shaping the intricacy and beauty of life.

It may be that much or all of my criticism can be answered, and that some explanations will be deeply embarrassing exposures of my misunderstandings.   But if I succeed only in causing those in the scientific community to explain themselves in terms understandable to the common man, you and I will have accomplished a mission.   But it is my hope that I am showing real holes in their fabric, which they will repair for the good of mankind and themselves.

Love

Dad

Herbert L. Nichols, Jr.

Contents

1. Public Events

During the past decade, there has been expression of public resentment at not being allowed to participate in decision-making of elected or appointed officials. This has resulted in various laws requiring all-public discussion of issues, and opening government files to public inspection. Their usefulness appears to be doubtful, to put it mildly, but that is not our subject.

In a wider field, vast numbers of the public participate in many events through television, radio, newspaper and magazine coverage. It would be reasonable to assume that scientific and engineering findings that become widely publicized would tend to be reviewed, criticised, and corrected, because of their exposure to thousands or millions of minds, both scientific and otherwise. But this does not seem to be the case.

In this chapter, examples are given of obvious error (charitably speaking) and of gross omission, exposed to vast, intensely interested audiences, in which little or no corrective response occurred.

Falling of a bridge

On June 28, 1983, a 100-foot long section of a three lane bridge on Interstate Route 95 (the Connecticut Turnpike) across the Mianus River, in Greenwich, Connecticut, a few miles from the New York State line, fell about 70 feet into the Mianus River. It took down trucks, cars, and people, three of whom died.

The fallen section (like three others in the same pair of bridges, all called suspended sections) was supported at one end by a pair of horizontal nine-inch steel pins, and at the other end by rollers. Pins and rollers transmitted weight to supporting or cantilever sections supported by concrete columns, per diagram. This allowed for heat expansion and contraction, and absorption of vibration.

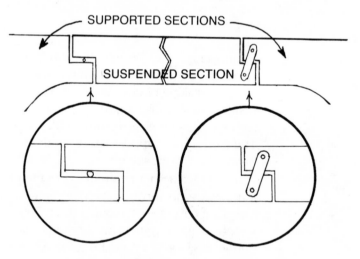

Free-Fall bridges, actual design

It also placed a great deal of dependence on the pins. If they broke, or worked out of engagement, their end of the section would fall. The rollers would then roll off their tracks, so that their end would drop also, only a fraction of a second behind.

But why fall so far?

The collapse caused an uproar, of course. Blame was placed on inspectors, inspection methods, the State for having deadlined an inspection gadget (the snooper), on pin and bracket design and strength, and the maintenance budget. Reports were often highly official, and included studies by prestigious outside engineering firms.

(I cannot resist a side remark that the peeling paint and rusting surfaces of the steel structure were a disgrace to any highway authority, and particularly to one generously supported by tolls. However, that is a different problem.)

Criticisms were offered by official examiners and consultants, and by a variety of experts and non-experts. Each is more or less valid, but NONE of them (except mine) dealt directly with or mentioned forcefully the only important deficiency of the bridge, that turned a maintenance problem into tragedy — the lack of secondary support, to hold up the span if the pins failed.

Such support could have been easily provided in any of several ways. Any substantial projection of the cantilever (pier-supported) beams under the suspended ones would have surely prevented the tragedy. See the illustrations.

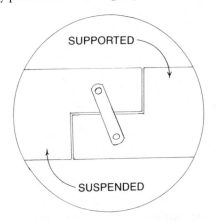

Would-have-been-safe design.

If such supporting projections were present, pin loss or breakage would merely cause the suspended section to sink a fraction of an inch. Pin replacement might be an awkward job, but could be done when convenient, as its lack would not prevent use of the bridge.

To reinforce an existing bridge, massive extensions can be welded or riveted to some (or all) of the cantilever beams, to extend under the ends of the beams in the pin-supported sections, per attached sketch.

Many other bridges now have the same built-in requirement to fall if pins fail. Their modification for secondary support should have URGENT rating in highway budgets.

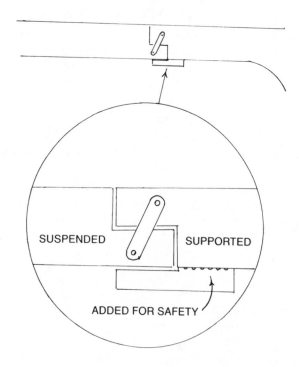

SUSPENDED   SUPPORTED

ADDED FOR SAFETY

Quick way to fix Free-Fall bridges

This could have been done at once with the other free-fall sections of the Mianus bridge, so that one side could have been opened to all traffic within a few days, instead of subjecting Greenwich (and for a while, Port Chester, New York) and turnpike traffic to the long agony (many weeks for trucks) of a closed bridge and a congested detour.

The section that went down was of very heavy construction, and fell into mud. It would seem likely that it could have been lifted back in place (quite a lift, 500 tons, but possible), and re-pinned, but it did not seem that this was considered. I think they cut it up and took it to a yard where they could look at the pieces, although only two corners were involved in the collapse.

There are sinister side situations that have come to the surface during the Mianus discussions. An engineering report on the failure mentioned, without great interest, that the design "lacked redundancy".

Following this clue, I found that among engineers, or at least bridge engineers, any secondary safety device, such as my beam projections, is redundant. The word is one of criticism and contempt, meaning, among other things, excessive, superfluous, and more-than-required.

The attitude behind choice of this word seems to indicate a colossal conceit. A routine policy of omitting back-up safety arrangements requires that these engineers have total trust in their designs, a childish faith in the competence and integrity of those who turn the designs into structures, and/or callousness toward the safety of the users of the structures.

But to return to my main point, that of public participation, I still find it hard to believe that among the hundreds of highway officials and engineers considering the disaster, and the millions of the interested public, I was the only one to point out publicly that it was lack of safeguards to prevent free-fall of the section, not undetected pin failure or other problems, that was the entire tragedy.

Back a few years

This is a bridge story, heard and believed long ago. It has a toe hold in this chapter only because there is another bridge in it.

From nearly their beginning, it had been customary to build steel suspension bridges with heavy vertical reinforcement at each side of the roadway, so that the bridge deck was rigid enough to resist localized distortion from traffic loads or wind gusts. At the time of the Brooklyn Bridge, around 1880, it was a necessary part of design.

But in the 1930's, engineers with sharp pencils went into the subject deeply, and figured out that such reinforcement was not needed, as the deck could readjust itself smoothly on the numerous vertical cables that hang it from the suspension cables. Steel and money could be saved by leaving it out, and the bridge would have a cleaner look. Vertical stiffening became redundant.

At least two bridges were designed and built this unstiffened way — the Tacoma Narrows span across Puget Sound in Washington State and the Bronx Whitestone across Long Island Sound (East River?)

... stiffening became redundant.

There was disagreement among the Bronx-Whitestone engineers, one of whom decided, rather late in the work, that leaving out the vertical stiffening was unsafe. He was unable to either convince his colleagues or reassure himself, so he resigned.

A while before the bridge opened, in April, 1939, this engineer made a concrete foundation in a then-existing garbage dump east of the Bronx toll booths, and used it to mount a powerful camera with a telescopic lens. He took pictures of the bridge under various wind conditions.

The pictures revealed the deck in excessive up and down and twisting movements, definitely greater than allowed for in the design. He showed them to responsible parties, and

6

was told to stop being a nuisance and a troublemaker.

The Tacoma Narrows bridge collapsed in a spectacular manner, on November 7, 1940, four months after its opening. A moderate (42 mile an hour) wind had gusts so timed that they made the deck pulse up and down in rhythm, and twist. The motions increased until it broke apart, and fell.

Almost immediately afterward heavy vertical reinforcement was installed on the Bronx-Whitestone deck, and the parallel, same-management Throgs Neck bridge, then planned or building, had it installed as it was built.

This story (whether entirely true or not) illustrates the danger of changing an old or standard design, particularly to make it weaker or lighter, or eliminating a part or procedure with security value, unless very extensive and reliable tests can be made. And it emphasizes the recklessness of ignoring warnings from knowledgeable people.

Both engineers and buyers have an obligation to insist on security. The word 'redundant' does not deserve a place in the conscientious engineer's vocabulary.

"From the moon"

There was a television broadcast of astronauts walking on the moon.

We saw them crossing a large, steep-sided gully, which then or later got the name, "Hadley's Rille". Its sides

Photo: NASA

seemed to show a series of thin layers of rock or soil, six to eighteen inches thick, with a slope generally similar to that of the gully floor.

On Earth, the gully shape would have indicated erosion by running water. The banded sidewalls looked like underwater beds, raised and tilted, or deposits of volcanic ash on the existing slope.

Each explanation is contradicted by our knowledge of the moon. We are quite sure it has never had enough liquid water to build sediments or carve gullies. We believe that there have never been mild, layer-building volcanoes.

This formation therefore aroused discussion and speculation, particularly as overhead pictures show the Rille to be hundreds of miles long.

A few days after the pictures were shown, it was announced that the Rille had been identified as a collapsed lava tunnel. But it has no resemblance whatever to a lava tunnel, collapsed or otherwise.

A lava tunnel is formed by massive leakage from a central volcanic duct, or drainage from the interior of a crusted flow. It tends to be round like a big pipe, and becomes lined with thick, hard lava, very black. It is a totally different type of feature from the gully-like rille, and it is something that quite a few people have seen.

Nobody argued.

In the aftermath of the splendid accomplishments of moon exploration, it is unimportant that a small problem is set aside under a doubtful label. It is just given as an illustration of the futility of depending on wide publicity to assure truth, or even thorough discussion.

2. Geology

Splitting continents

An oddity of Science (the scientific world) is that it will seldom accept a fact, however well proved, or even be particularly interested in it, unless (or until) there is a good
theory to explain it. This oddity is of course grossly
unscientific, and it has led to embarrassments.

An example is continental drift. In 1912 Alfred Wegener, a German meteorologist, was so impressed by the
apparent fit of South America and Africa (bulge into hol-
low), that he decided that they had been part of one conti-
nent (Gondawanaland) which had split, and the parts had
drifted apart, leaving (or creating) the Atlantic Ocean to
separate them.

He and others made detailed examinations of rock formations on the two sides. It was found that distinctive strata
in Brazil were found to reappear in Ghana in Africa, across
thousands of miles of ocean. Recognizable rock formations leading into the water in Nova Scotia were found to
reappear in Scotland.

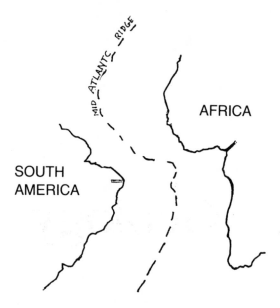

In each case the continuity of the formations was shown positively by rock type, bedding, and crystalline details. They provided nearly absolute proof that continents on opposite sides of the Atlantic had once been joined.

Anyone with a bit of knowledge of geology, and money to travel, could see this for himself. But leading geologists doubted this FACT, or at least thought it not worth serious consideration, on the ground that Wegener's THEORY of what made the continents split and drift was unsatisfactory. His facts and his theory were lumped together, and totally rejected. They were considered offbeat and visionary, just barely allowable in idle and unofficial talk.

Now there is a new theory, plate tectonics, which accounts satisfactorily for the separation of the Americas and Africa-Europe. The generally north-south Atlantic Ridge, has been found to be building new ocean floor by upwelling of lava, as those floors move away from it, east and west, carrying the continents with them.

This theory and its supporting facts explain the splitting and movement of the continents, so that Wegener's continental separation and indicator rock beds were accepted. But he, poor man, was dead.

But he, poor man, was dead.

It would of course been far more scientific (in its true sense) to accept the fact of separation, theory or not.

Settling this problem aggravates another. It confirms an Atlantic age of over 150 million years, while European and American life forms show quite recent connection.

Volcanic violence

The spectacular and destructive behavior of volcanoes causes them to be extensively reported and discussed in newspapers and magazines, often by lay writers or by scientists specializing in other fields. The majority of the articles follow an old tradition, or perhaps superstition, that explosive volcanic activity is largely dependent on the nearby presence of large amounts of surface water.

In support of this idea, it is pointed out that a large share of the earth's volcanic activity occurs in a "Ring of Fire", a chain of volcanoes stretching most of the way around the Pacific Ocean. It is a significant appearance on a small scale map, but many of these volcanoes are far from the shore, up to hundreds of miles inland.

Most active volcanoes are found in areas of crustal disturbance, where crustal plates are pushing or rubbing against each other to form mountains, or moving away from each other in oceanic rifts.

In general, a volcano is a surface outlet for an underground pool or lens of melted rock, which may have temperatures between 800 and 1500° Celsius (1500 to 2700 F.), and is slightly lighter and bulkier than the rock from which it was formed. The fluid rock is called magma while underground, and lava when it has reached the surface.

After cooling, magma is basalt rock, while lava may then be called basalt, or lava, pumice, cinders, or ash.

Temperature in the earth increases with depth, at a rate that varies greatly with locality. Melting point of rock may be reached a few miles down in volcanic areas, or at a possible depth of 60 to 120 miles elsewhere. But it usually does not melt, because it would have to expand to do so. Weight of rock above it, and the tightness of rock around it, keep it squeezed into a solid state.

If the weight and pressure are reduced, by an upward thrust from a crowding plate or local movement, or by the pulling apart of crust, a mass of rock can melt, and form a pool or reservoir. This relief of pressure or weight seems to be the start of most volcanic action. It is typical of ocean rifts, where the movement of crustal plates away from each other is accompanied by an almost continuous line of volcanic action, sometimes thousands of miles long.

Rock usually contains combined or dissolved gases, often in very large amounts. The most potent of these is water, which at very high temperature and/or pressure may separate into hydrogen and oxygen, which combine peacefully with chemicals in the rock. Slight cooling (I believe to 900° Celsius) re-forms the water as superheated steam, which will expand explosively if released.

The molten rock might work its way upward, by pressure-opening cracks, and/or melting holes, until it reaches the surface. Or it may be blocked part way up, not ever showing us anything but hot springs and a bit of steam.

If successful, it may have forced its way through a great deal of more or less intact rock if it is a new volcano, or through thoroughly plugged passages in an old, dormant one. Fluid pressure at the top may remain nearly as great as in its reservoir, squeezed by miles of overlying rock.

Once a breakthrough is made, rush into open air of the top of the magma (now lava, cinders and ash) reduces weight on and pressure in the whole fluid column. Steam and gases are released, resulting in explosion or explosive ejection of the gas-lava mixture, like opening a bottle of warm, well-shaken champagne or soda, but on a vast scale.

If the magma were thin and/or the passage open, steam and gases could have been separating during much of the upward passage, and might escape into the atmosphere with little harm (or anyhow, much less), leaving the lava to flow out more or less peacefully. This is usual in oceanic volcanoes, but even they may explode occasionally.

Due to variability in magma fluidity, gas content, pressure and temperature, and circumstances in the outlet, almost any combination of these effects may be produced. However, they are usually all parts of the internal mechanism of the volcano, and have little reference to surface characteristics.

But in popular press articles, it is commonplace to find violence in an eruption blamed on the presence of nearby bodies of water, or high water content in local rocks. Any effect this might have is trivial compared to the power of the gas-steam expansion from confinement.

Articles on ocean spreading seem to consistently ignore the pressure-relief factor of the separating crusts in causing volcanic action. In fact, it is occasionally debated whether magma pressure against the slab edges is the power that moves them apart.

A different mechanism creates some ocean-island volcanoes. They seem to arise from a very deep pressure area or plume, from which the magma forces its way up through the whole thickness of the crust slab. Movement of the crust over the fixed plume creates a chain of volcanic islands, only the most recent of which is active.

The Hawaiian Islands are a prime example. They seem to indicate that the Pacific plate has been moving easterly and a little southerly for some time. There is an almost-connection with the Emperor Seamount Chain, extending northerly to the Aleutians.

To summarize, there is probably enough earth heat any-where to fuel a volcano, but limited depth, with crustal movement to reduce pressure at that depth, are usually necessary to create one. Violence of eruptions is due to release-expansion of steam and gas confined in the magma, and is seldom related to the presence of surface water.

3. Cosmology

Cosmology, also known as astrophysics, is the study of the universe, which is sometimes called the cosmos. Cosmology relies on astronomy and nuclear physics for most of its data, but its interpretations involve philosophy, mysticism, mathematics, and imagination. It is not generally considered to be a science.

Cosmology operates on the slenderest threads of evidence of any of the explorers of physical phenomena. Much of their crucial material is so far away that it is barely recorded by the most powerful instruments. Yet their reports and theories are often or perhaps usually presented with complete confidence. Qualifying words such as "perhaps", "maybe", or "possibly" tend to be quite rare.

There was an executive above me in my youth, of whom we said fondly, "He is often wrong, but never in doubt."

Astronomy and cosmology operate in a universe of vast distances, and we have had to develop special measurements for them. Biologists may (and nuclear physicists must) work in a world of the almost incredibly small, and need appropriate measurements also.

A description of the overall measurement system follows, for a couple of pages. If you are already familiar with it, or don't care about it, it will do no harm to my story to skip past.

For space, we start modestly with an astronomical unit (a.u. or au), the average distance between the Earth and the Sun, about 93,000,000 miles or 149,600,000 kilometers. The planet Pluto is about 80 au away from us.

The light year is the standard unit of measurement for most cosmological distances. It is the calculated distance that light would travel in a vacuum in one year, moving at 186,000 miles (300,000 kilometers) per second. It is about 5.878×10^{12} miles, or 9.46×10^{12} kilometers.

Way-out numbers, either tremendous or tiny, are figured in powers of 10, as shown by an exponent figure. Exponents are written in superscript, meaning above the bottom line, and should be a smaller type size than the text, as the ^{12} in the paragraph above.

... in powers of ten...

The ^{12} is the twelfth power of ten, meaning that it is ten multiplied by itself twelve times. Written out, it is a figure one, followed by 12 zeroes. Translation into ordinary figures is easy, just write down the digit 1, and then as many zeroes as indicated. Our 10^{12} works out to 1,000,000,000,000, or a million million, taking lots more space than it did as a power. To go the other way, count zeroes and use their total as an exponent.

On more ordinary scales, 10^2 is 100, 10^5 is 100,000. The 10 and its exponent (power) figure is followed by the unit of measurement; kilometer, light year, or whatever.

An exponent with a minus sign, such as 10^{-3}, means a fraction — a fractional power — with the basic 10 divided by 10, the number of times indicated. The fraction has the digit 1 as its numerator, above the line. The denominator, below the line, is a 1 followed by the number of zeros required by the exponent. Our 10^{-3} is 1/1,000.

Powers are almost always of 10, to simplify arithmetic. They take the problem or statement up or down to its size area, in which it can be multiplied by ordinary numbers. We might have 2.32×10^7, which would be $2.32 \times 10,000,000$, to equal $23,200,000$. It could be written 232×10^5 for the same result, but a single figure before the decimal, with the higher exponent (power) is much more usual.

Small measurement units — shorter than meters — fit directly into the metric scale, as they were set either when the scale was worked out, or afterward.

The standard unit of the fractional parts is the millimeter, one one-thousandth of a meter, 10^{-3} or $1/1,000$. It is used for machine part sizes, paper thickness, and precise measurements of things around us, in general.

Next down is the micron or micrometer (but a micrometer is also a measuring device), 10^{-6} or one-millionth of a meter. Then the nanometer, 10^{-9}, the angstrom, 10^{-10}, which is handy for measuring atoms, the picometer, 10^{-12}, and finally the fermi, 10^{-15}, for nuclear particles.

There are two other fractions; 10^{-1}, the decimeter or $1/10$ meter and the centimeter, 10^{-2} or $1/100$ of a meter, around ⅜ of an inch.

The meter and the last two fractions are the ONLY parts of the linear metric scale convenient for ordinary human, household use. The fractions have nevertheless (or perhaps therefore) been practically dropped from the latest metric standards, the Systeme Internationale.

Such callous treatment of human use and value is probably a principal factor in the failure of the recent government-backed drive to make U.S. go metric.

Most cosmologists agree that the universe was formed by the titanic explosion (Big Bang) ten to twenty billion years ago of a body that contained all the matter that there is. Our universe, empty before the explosion, is now made up of this material, largely gas and dust, but partly assembled into the suns (stars) and other bodies.

All of it is said to be outward bound with the force of the explosion. The furthest-out parts are going (receding) the fastest, as viewed or measured from Earth. Top speed is about 80% of the speed of light, and it is expected that improved instruments will be able to detect further-out objects going still faster.

Speeds closer to us are less. Distribution in space is found to be fairly even although lumpy. Most stars are gathered into formations called galaxies, some galaxies are clustered in groups, and there seem to be empty areas.

The expansion of the universe is indicated and measured by light shift. If a source of light, such as a star, is moving away from us, its spectral lines, which are individual for each element, will appear more reddish in color than normal. This is called a redshift. If it is moving toward us, the shift will be toward blue. The amount of shift is proportional to speed.

Light shift is comparable to change in pitch of sound when the emitting object moves rapidly toward or away from the hearer.

Redshift is convincing evidence of expansion of the universe. Moderate distances to nebulae (other galaxies) can be measured by means of certain stars that are recognizable by variation pattern or other means, and that are all of the same brightness. Their distance can be figured by their luminosity — the further the dimmer. It usually (maybe always) agrees with the redshift calculation. Also, using the width of the earth's orbit as a base line

... like dust in a still room...

(observations six months apart), tiny differences in direction make it possible to figure quite long distances. This is called taking the parallax of an object.

The big bang theory is also said to be proved by the existence of microwave (thermal) radiation from all directions in space, at a temperature of 3° Kelvin. That is three degrees Celsius above absolute zero. Also minus 270°

Celsius and minus 457° Fahrenheit.

I have the deepest respect for the men and methods that detected, isolated and measured this incredibly faint radiation. But I have seen no explanation of a way in which it (traveling at the speed of light) could have hung around us for fifteen billion years or so, like dust in a still room, waiting to be measured. It would seem

... really pushing it.

more reasonable to consider it to be the average background temperature of our present universe, averaging stars, dust, gas, and emptiness.

Convinced and relaxed by redshift, and reassured somehow by background radiation, both cosmologists and astronomers accept an expanding universe as a basic fact, and most of them like it with an explosive beginning.

But one group denies the explosion, and says that space itself is expanding, so that stars move away from each other like raisins in rising bread, a theory that has problems too.

But as to the big bang itself, the people dealing with it seem to have pretty well forgotten how unlikely it is; how contradictory of good science and reasonable thinking.

As a beginning, they have no concept — not the faintest — of how such a mass could have packed together, nor how gravity could have then released to let it go.

The description of this mass that exploded to yield the big bang is variable. Originally it was all the mass of the present universe, compressed with inconceivable force into a "small" lump, perhaps a few light years in diameter.

But even this modest size caused difficulty to cosmologists, who wanted to get things moving real fast. Several of them figured that the nature of the universe was determined in the first 10^{-32} or 10^{-35} second (10^{-35} is 1/100,000,000,000,000,000,000,000,000,000,000,000) of the explosion. But it worked out that even at the speed of light an explosion couldn't go through any reasonable mass that fast, and they just had to have their 10^{-35}.

So now (at least for early 1984) the original mass for explosion has been reduced to a point (a point has zero area and volume) of infinite (even though infinite is a dangerous concept to play with in science) mass. The too-slow action is eliminated, but at the cost of introducing a special creation of an unimaginable substance.

It is a basic field rule of science to avoid, or at least attempt to avoid, theories based on once-only and unlikely-to-be-repeated happenings. In this regard, it seems bad enough to be working with a big bang at all, so decorating it with special creation of special impossibly dense material in a point, is really pushing it.

As a slight side trip, the most rational explanation ever offered for the formation of our solar system is that a wandering star passed the sun closely enough so that they drew material out of each other, much of which remained in space after their full separation. Such material could have worked itself into planets more readily than if from any other source, such as the now-popular condensing nebula.

This theory, advanced in the twenties as the Chamberlin-Moulton hypothesis, enjoyed wide acceptance until recently. Then, Scientific Opinion scrapped it because they couldn't locate the other star, and the hypothesis called for a possible-once-only happening. This change of attitude was at about the time that the Big Bang won acceptance.

Perhaps two rule violations at once were too much, so the less exciting one had to go. But

Have we a right to laugh?

more likely it was because an enthusiasm had developed at that time for life on millions of planets around other suns, and you can supply more planets with life possibilities using nebulae than with chance bypassing.

Anyhow, Big Bang has been reconstituted to depend on special creation and a special state of matter. While these may be conceivable to some, they are not science, and work with them is pure speculation. It is too bad that these theorists had to be in such a hurry inside that first second.

It is said that medieval theologists would argue lengthily about the number of angels that could dance on the head of a pin. Have we a right to laugh?

A visitor asked a hillbilly why he had three holes in his door.
He answered, "Cause I have three cats."
Question, "So what? Can't they go through one hole one at a time?
Answer: "Nope. When I say 'scat' I mean 'SCAT'."

Turning back to a real explosion of real matter, we must face the question of how it got matter so evenly (although lumpily) distributed in space, just at the time we got smart enough to look at it. An explosion tends to create a shell of flying pieces, leaving just smoke and fragments too big to move in the interior.

Where was that bang? Its backers seem to have little interest in that. They have even said — in print — that since it was all the material in the universe, it occurred every-where in the universe. That is nonsense, or perhaps only metaphysics, although they claim that it is according to

Einstein. An unconfined explosion expands from a point or area, throughout a larger space. If it occurred everywhere in space, it was not an explosion.

The same limitation applies to material introduced through or by a point. The point has to be at a specific location. We should be able to locate that spot unless it is very close to us.

Using redshift as a measure, the apparent expansion rate of our universe is very regular. Each galaxy or space area is going faster than the one behind it, which is closer to the explosion center. Likewise, the next one further out is going faster than it, and so forth to the end. We are leaving the next neighbor behind us at the same rate the one ahead is leaving us. There is an increasing speed and distance between each pair, with mounting speeds as far as we can see, up to 80% of the speed of light.

The regularity of this spacing system should awaken

Engineered for uniformity

suspicion, especially because it works equally well backward and forward, and in all directions, as if we were at the explosion (bang) center, and that the bang had been engineered for uniformity at the exact time we started looking through telescopes.

If there was a bang, and we are not near its center, we should be able to determine the location where it happened by fairly simple methods.

The illustration shows a sketch of a narrow wedge view of part of the universe, extending from the supposed explosion point outward. Three spokes are arranged to show a few galaxies on radial lines from the center.

If we are at point (or galaxy) A, the galaxies B and C will have red-shifted light to indicate recession from us, at approximately equal rates as mentioned above. The cosmologists assure us that this is somehow a normal result of looking backward and forward along our own line of big bang propulsion.

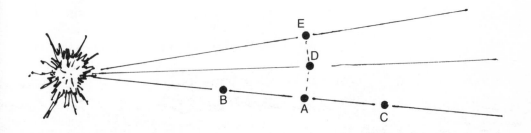

The galaxies D and E have distances from the center similar to ours, and are moving outward on lines nearly parallel to our own, at speeds that should be nearly the same as ours.

As a result of similar speed and nearly parallel movement, speed of recession from us should be very slight.

Therefore, we should make observations of neighboring galaxies in all directions, estimating their distance by brightness of certain variable stars, and measuring their redshift. If we are really speeding outward from an explosion center, there will be substantial redshift ahead and behind, and little to the sides. From this data, the center could be located by surveying methods.

If a galaxy, or a group of them, should show very high redshift, double or more the greatest expected at their distances, we might assume that they were the other side of the center, and outward bound the other way.

Since the principle involved in this checking is so simple, failure to do it, and to publish results, seems to show a possibility that the big bang people are not convinced themselves, and would prefer to avoid a test. Or — sobering thought — they may really think that the explosion happened everywhere.

Redshift is the only real evidence for an expanding universe. If (I wish I dared to say, "When...") it is found that light can get tired, and redshift may also be caused by distance only; the Big Bang will be gone, and the universe studied on a more rational basis.

Big bang probably comes first in the interest of the astrophysicist or cosmologist, but gravity is a close second.

Gravity is a powerful force, but so simple that it is not understood at all. It is an attraction between any two units of material that have weight, which are called masses. (Mass equals weight). Its power is in proportion to the product of their masses multiplied together, divided by the square of the distance between their centers. Cosmologists tend to overlook the square-of-the-distance factor, which reduces remote pull very severely.

In our understanding of the cosmos, gravity has many jobs. It heats new suns to the nuclear fusion point as their material sinks toward their centers, gathers groups of galaxies together, gives spiral galaxies their shape, handles complicated exchanges of material in pairs of stars; all in addition to its basic job (to us) of keeping Earth and the other planets revolving around the sun.

There is said to be no limit to its reach. A star remote and dim in the sky is said to be pulling upward on you and on the ground you stand on, but so weakly that the force cannot be measured.

Our modern cosmologists believe that faint as it is, the gravity pull of star on star is moving them, and grouping them into clusters, galaxies, and clusters of galaxies. (Our galaxy is the Milky Way, and everything inside it — billions of stars, with endless clouds of dust and gas).

Many cosmologists further believe that these faint pulls combine to form a drag on the outward speeding fringes of the universe, and will eventually stop them and pull them back, first slowly, then faster and faster, until they unite and are ready to explode again.

This theory is usually called that of the closed or pulsating universe. The open universe theory says that gravity is too weak to stop expansion, which will go on forever. And there is a minority that says that it is not expanding at all.

The two yes-no groups try to settle their dispute by weighing the universe (don't ask me how) to find whether there is enough mass in it to create enough gravity to close. This mass has just been given a big boost by finding vast and heavy clouds of non-luminous dust.

As we will soon see, this problem probably does not exist. But if it did, the trouble of weighing the universe could be saved by working out a different approach on the back of an envelope. If we had a big bang 15 (or even 10) billion years ago, the fringe is still receding at about 80% of the speed of light, which is the maximum possible starting speed, as scientists agree that nothing can move faster than light. It may not have slowed at all, it can't have slowed more than 20%, in about 15 billion years and a lot of light years. This distance, so vast as to be inconceivable, must be squared when figuring reduction in gravitation.

If we concede that the universe is expanding at this rate, we have a clear case of escape velocity. There can be no falling back, whatever weight the universe may have.

But the dispute is an empty one. Cosmologists assume that every star pulls on every star, effectively. They think this is proved by groupings — planets around the sun, galaxies around a central core, even groups of galaxies. If these are not bound by gravity, how can they exist?

But careful observation inside these groupings, except close ones like sun and planets, tell a different tale. There are strong indications that there is no effective gravitational attraction between stars at ordinary interstellar distances, whether they are quite separate individuals, or are in clumps or masses.

A typical spiral galaxy appears to revolve around a single axis, like a vast, thin disc, bulging at the axis.

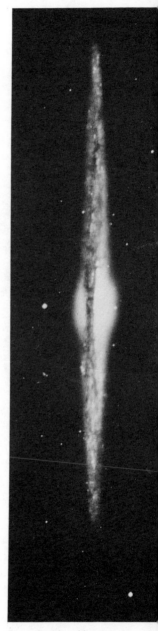

Photo: American Museum
of Natural History

The assumed (but difficult to check because it is so slow) motion in these bulges is revolution around the axis, in common with all other matter making up the galaxy.

Photo: American Museum of Natural History

If star were attracted effectively by star, those in the bulges would be pulled into the central plane of the galaxy. Axial rotation would give no resistance to this side movement. But they often do not seem to be. The axial bulge is a common feature in the spiral galaxies we can observe edgewise.

In our own galaxy, and in at least some others, the bulge contains many globular star clusters. They are spherical or almost so, and are made up of perhaps millions of rather closely spaced stars, many or most of which are old.

So far, it has not been possible to detect any orbital or regular movement inside the clusters. Nor has it been possible, even with computers, to calculate any orbits for stars in a spherical group that would prevent them from settling into each other, if gravity were pulling them.

The close and regular groupings of stars in bulges and clusters certainly looks like assemblage by gravity. But by holding their positions against further inward movement, they seem to deny the existence of any attraction that could collapse either their formations or the universe.

The stars and material in the spiral arms of galaxies are generally assumed to be gravitationally bound to the center, so that their rotation around it is planet-like. But their speed is known to be wrong, and even when corrected with new heavy dust clouds bordering the galaxy rim, it is still wrong.

If star were attracted by star...

A slight possibility is that arms are being spun off the core by centrifugal force, from some source that provides material for the arms and the bulges. But what could be the source supplying each of millions of galaxies?

The question of gravitational action or non-action among stars deserves more careful and open-minded study.

4. Evolution

Evolution is a classic example of the problems, blunders, and random hostility that may surround and pervert a scientific discipline.

The theory or science of evolution is that life on Earth originated in chemical reactions over a billion years ago; that the first forms of life were very simple, that the simple forms gave rise or birth to more complex forms, and that practically all life now on earth has descended from an original humble ancestor.

So far, the story is reasonable and the evidence for it is overwhelming. Geologists have found at least a few samples of sedimentary rocks from almost every period of pre-history. There are at least three ways to find their approximate times of formation.

First, position. Newer rocks are on top of older rocks, except in a few places blocks of layers of rock have been turned on their sides or even upside down, a situation that causes only temporary confusion. And almost from the beginning of geology, many rock formations could be identified as to age by comparison of contained fossils (mostly preserved-in-stone or turned-to-stone parts or casts, of living things) with similar or identical fossils in other formations, perhaps in the next county or even half-way around the world.

Third, it is possible to date many of the rocks by the stage of decay of radioactive materials included in them when they were formed.

Recently, it has become possible to analyze the contents of living cells, and particularly of the genes that determine their form and activity, enough to determine the closeness of relationship between the organisms from which the cells were taken. In general, results agree with the fossil record, and confirm the single-ancestor idea.

It is reasonable to say that the fact of evolution is proved, and that we, if rational beings, must accept it as a fact.

But the theory of evolution — the explanation of these facts — is another thing. It was set up in 1859 when Charles Darwin published his ORIGIN OF SPECIES. On the basis of thoughtful observation and a large measure of inspiration, he declared that evolution depended on just two factors: heritable variations constantly arising in individuals, and natural selection eliminating variations or original structures that made survival of individuals less likely.

... evolution is proved...

The later establishment of Mendel's laws of inheritance, published in 1865 but not recognized until around 1900, were absorbed into Darwinian theory by a currently still-dominant group called the synthetic school, "synthetic" because they blended the old (1859) with the newly found genetic mechanisms.

At present (1984) Darwin's evolution theory (Darwinism) is dominant and substantially unchanged, and rests (1) on heritable changes in organisms caused by mostly-random changes (mutations) in content or location of the genes that dictate their structure, and (2) natural selection determining the ability of organisms, changed or unchanged, to survive and reproduce.

There are many variations and arguments within this structure. One of the latest is a macro or jump procedure, in which important and perhaps revolutionary changes appear in genes or groups of genes, abruptly, instead of smaller ones frequently and more smoothly.

No mechanism seems to be offered, and the long range results would be similar, and suffer from the same questions of cause.

Another recent arrival is the selfish gene. It is argued that the body, whether of man or dandelion, is merely a gene's way of providing for its own survival. This is a finer tuned version of the old saying that a chicken is merely an egg's way of making more eggs.

... mock the whole system.

Thus any traits that do not clearly contribute to the body's (and thereby the gene's) reproduction, will be pruned off very sternly. Social, cooperative, altruistic or even just neutral traits have to be reviewed by researchers, who patch them in, usually insecurely and with difficulty. This struggle seems to be the principal activity in evolutionary theory at present.

Selfish genes seem superficially reasonable, but exaggerate and even mock the whole system. They present some interesting twists. However, they are a side issue.

At its beginning, Darwinism had a mixed reception ranging from enthusiasm to condemnation. It was generally respected as a profound and brilliant concept, but objections were found.

The scientific world accommodated other theories also. One was the older Lamarckian system, that based evolution on the direct inheritance of acquired characteristics. For example, if an animal grew extra warm fur in a cold climate, or strong leg muscles from lots of running, it would pass these assets on to its young, partly or wholly.

This theory was most attractive, as it accounted for the usually excellent adaptation of organisms to the circumstances of their lives. But researchers could not produce such results experimentally, and technical problems prevented formation of a good theory of how Nature could do it. As a result, the idea was mostly abandoned early in this century (except for a long revival in Russia inspired by T.D. Lysenko), and only a few die-hards fight for it now.

Orthogenesis was another popular theory. In this, the hereditary mechanism (first plasma, then genes after they were discovered) might have a compulsion that made most or all mutations change a species in one direction, which might or might not be good for efficiency and survival. An extreme example was the Irish elk, whose antlers kept getting bigger even after they were too big to be managed. They are believed to have been a major factor in extinction of that species.

This theory was also attractive, as it made evolution seem more purposeful than Darwin's random mutations. But here too, no good evidence could be found in either laboratory or field studies that there was ever bias or direction in mutations, nor could any mechanism be found that could create and regulate such bias.

Creationism — total reliance on the Biblical account of creation — does not seem to have been particularly active during this period. However, there were carefully considered theories in which the physical mechanism of evolution was conceded, but in which it was regulated by divine purpose and power, either in the environment or in the hereditary material. But these approaches did not work.

But researchers could not produce such results

In the twenties, the Darwinists joined forces with the Mendelian group in studying the mechanism of inheritance through the separate and distinct paired packets of controlling units called genes, creating the Synthetic School of Darwinism.

From then until now, these people have been making studies and writing books. They have collected much valuable information, both in regard to the complications of gene behavior, and the related shaping of organisms by natural selection, and have published many detailed reports and commentaries.

However, few or none of the basic objections that had originally made natural selection unconvincing were dealt with at all. It seemed as if the Synthetic School, when asked an awkward question, would simply talk confidently and loudly about something else. And somehow, this worked. In their hands, natural selection rose from a marginal or almost discredited status to thorough acceptance.

It now appears that any article about evolution, in either scientific or popular press, simply assumes that natural selection is the one and only driving force. Discussion of-a particular trait may ask what its selective advantage is, but seldom whether there is any.

Agreement of scientists is not total, but the disbelievers are a minority. And, strangely, they limit their inquiries or protests to details or fringe issues. These people, along with the Creationists, seem held back by good manners or some internal censorship, so that they do not attack natural selection in any of its several very weak points. Or perhaps the attacks are made, and it is magazine editors who think they are in bad taste and not suitable for their readers.

One of the origins of this book was twenty-plus years ago, when I read a recently-published book by a leader in the Synthetic School of evolution. It was crammed with forceful arguments that promoted and defended the concept of natural selection, which by then had been so well accepted as the driving force of evolution that there was nothing to argue about.

It was as if a business man had only polite interest in his wife's daytime activities, and she in relating them. But she might start to be very specific about them — shopping, yoga club, computer classes, or whatever — with defensive remarks about dullness of just sittting home, continued diligence in housework, the moderate cost of her new hobby, and so forth.

An astute man might soon decide that she was covering up something about which she felt guilty. It would seem that an affair with a man with free afternoons would be indicated.

Only respect for his wife's privacy would prevent a checkup.

But the book (or rather, its author) was more difficult. It was obvious that he was covering up something. But what? And why?

It was a long reach to the realization that natural selection itself was the man in the case. Such absurdly exaggerated promotion and defense of it must have been caused by knowledge, perhaps subconscious, that it was not the answer.

But then, why champion it? To sell a book? Unlikely, in view of the stature and knowledge of this man — he could write about something else.

It must be fear of emptiness — trying to push away or bury the knowledge that here was something — something vast and important, strange and beautiful — that he did not understand at all.

The theory of evolution encounters more and/or stronger popular hostility than other branches of science that are equally Godless. This may be partly because we do not want animals as ancestors and relatives.

Another possibility is reaction to evolution's failure to account for or usually even consider the beauty of organisms, and their almost-orderly ways of life. Not consciously, but deep down, man may feel that Creation can account for the beauty and the fit of life, while an all-selfish competition fueled by blind chance cannot.

Criticisms of the theory of evolution are almost sure to be taken up by Creationists, to favor their story and their doctrine. Let us therefore look at what they have.

The writer of the account of creation in Genesis set down the myth (or myths) that were the belief of a half-savage tribe. He knew nothing about the nature of the universe, thinking that ocean water was lying in the sky, above the stars, and that light and dark were tangible things, separable like stone and water. He even has two versions of Creation, the second making mankind more important than the first. Both were in the sacred traditions, and must be kept, contradictory or not.

Then there is the Flood. During it, Noah and his family were able to maintain at least a pair of every land animal and bird species. We know that there had to be several million of them. They, and food for them for many months, were on a barge of limited size.

There is no mention of Divine assistance with the housekeeping. During this time, the whole earth was under water, all other people and animals were drowned, "...every living substance was destroyed that was on the face of the ground." But plants survived, apparently unharmed. Can we explain this to school children?

I would like to make a fatherly suggestion to the Creationists—that they abandon their campaign. They must know that they are wrong, and if they did somehow succeed, success would destroy them. Their basic interest is in defending and promoting a highly conservative (Fundamentalist) type of Christianity, that maintains the literal truth of every word in the Bible. In this setting, beautiful tales such as Creation and the Flood may continue to be accepted as a background needing no explaining, even though the facts of creation and evolution are also learned, and believed in a different way.

But they ask that Biblical stories be set up in direct classroom competition with the scientific theories, and expanded into full school courses. Then these stories, and the Old Testament in general, will be held up conspicuously to searching curiosity of youth. Students will ask questions, their own and those suggested by friends in competing courses. For many of the questions, there are no answers. Then the fiction that the Bible is all true, and the exact Word of God, must crumble, and faith be lost to children, as a result of a pretence of being science.

Here are two of the many questions. Why do we read in Genesis 15-18, the Lord's *covenant* with Abraham, that he has given to Abraham's descendants (the Hebrews) the land from the Nile to the Euphrates? Then in Genesis 48-4 we find that the land of Canaan is confirmed to them for an everlasting possession. But that has not been its history.

In Ezekiel 26-7 to 12, Nebuchadrezzar is promised that he will take and destroy the city of Tyre; in Ezekiel 29-18 to 20 Nebuchadrezzar is told that since he could not take Tyre, he can conquer Egypt instead. (But he didn't.)

Ezekiel did no worse than most of our own prophets. But for him our standards are strict; we are looking for Divine inspiration, and finding no trace of it.

In discussing the Old Testament, I cannot resist mentioning my moral revulsion inspired by Deuteronomy 23-1 and 2.

As in most mature disciplines, a rigid evolutionary code has developed. It includes worshipful respect for Darwin, and often fanatic belief in his ideas, together with whatever modifying ideas have been accepted. Faced with a new discovery, Darwinians do not ask first if it is true, but if it fits in their code. Speculation about what Darwin himself might have said about it is a major factor in acceptance.

A weight of unanswered questions increases as new instruments and techniques reveal almost limitless complexities (in the billions) in organisms, in both structure and function, but somehow the importance of this as a problem does not seem to be realized.

To repeat, the fact that evolution has occurred is clear. In a long time, well over a billion years, life has evolved from extreme simplicity to bewildering complexity. The evidences are too positive and too varied to be denied by any person capable of clear thought. They include the long sequence of the fossils, the succession of ancient to modern rocks in which they are found, the radioactive minerals and structure details that show their age, and submicroscopic relations in structure and chemistry that exist now among living species.

Hamlet:
"The lady doth protest too much, methinks!"

But theories of evolution are in deep trouble, not in arguments with Creationists, but in their fervent, almost religious belief in natural selection as the only force that shapes it. No article, and few pages, about either evolution or biology, fail to mention natural selection reverently at least once. Several times a page is usual. It is automatically credited with all organic change.

Long ago, when early scientists were becoming aware of how our world is made, many of them, in revealing their findings, paused over and over again to make comments such as, "Thus is revealed the wonder and glory of the Creator!" or "Such is the power and wisdom of God."

Such insertions often served the practical purpose of diminishing the dangerous hostility of the Church's fixed ideas, and we have no reason to believe that they were entirely insincere. But frequent repetition of this kind shows basic unsureness. (Hamlet: "The lady doth protest too much, methinks.") They believed in the truth of what they wrote; but did not dare to admit, even to themselves, how often it contradicted teachings of the Church, in which they still wanted to believe.

Similarly, our modern biologists tend to sprinkle their writings with their obeisances to Natural Selection, seeming to the critic to show a similar uncertainty. Motives are the same—to tangle with Natural Selection is likely to bring scorn and scientific eclipse. They have no theory to take its place, but common sense and intuition are unsatisfied.

It cannot be denied that natural selection has an important, a very important part, in evolution. It has an absolute veto. No life form, no matter how cunningly, ruthlessly or beautifully designed, can have a place in the world if it and its descendants cannot survive.

But with only random mutations in simple organisms to work on, natural selection cannot have built or even closely directed the building of the complex organisms and associations of life today. It is time that Science faced this problem directly, and intensified efforts to find other factors. The energy now wasted trying to force all characteristics into a natural selection origin might be saved, and severe future embarrassment avoided.

Study of recent literature, particularly scientific reports on biology and evolution, combined with observation of life around us, seems to indicate that the traits of organisms in relation to natural selection can be very roughly divided into three classes.

The first, about forty percent of traits considered and observed, conform clearly to natural selection precepts. Animals in cold areas have warm fur, meat-eaters have sharp teeth, hawks hunting small and remote prey have excellent eyesight, and so forth.

The second group, about neutral, with possibilities of force-finding selective connections, is also about forty per cent. This includes such items as trees' bark patterns and leaf shapes, sociability of animals and birds, colors on a night-flying moth, and music appreciation in a whale.

The last twenty per cent is clearly anti-selective. It includes relatively minor matters such as poisoned or bitter fruit, seeds too thickly armored to germinate normally, and poor eyesight. These are mentioned just for openers.

The more serious doubts about natural selection will be considered soon. But it is a curious thing that with the limited importance shown so far, it ever attained its present dominance in evolution theory. The answer must be that it is because there is nothing—absolutely nothing—to replace it, as all the other approaches have been shown to be even weaker. Scientists, even more than Nature, abhor a vacuum. They will not, seemingly cannot, face an emptiness and admit that they do not know.

They offer no explanation for beauty, except in the many cases where it serves in sexual selection, or is a byproduct of protective coloration or mechanical efficiency. In much of the organic world, it is of a quality and an extent that far exceed these limited purposes.

Still in the twilight zone, evolutionists (or, more accurately, "natural selectionists") have failed to cope with, or even publicly recognize, a life characteristic that may be called "fit", or less completely, "interspecies adjustment."

In general, an ecosystem, which is an assemblage of a variety of interacting life forms in a limited area or habitat, involves many adjustments among its life forms, from complex to simple, that cannot be explained by natural selection, particularly with the selfish gene approach.

The balanced ecosystem is good...

For example, in a steady or stable system, animal birth rates tend to be just about high enough to maintain a species, making up for normal losses by disease, aging, predation, accident or whatever. A mouse, short lived and exposed to untimely death on many fronts, may have two litters a year, with a dozen mouslings in each. An elephant, long lived and secure, may give birth to one calf (or rarely, two) every two to four years. Most birth rates lie between these extremes.

Their friends keep saying in defense of selfish genes that lower birth rates merely make possible better and more devoted care, hence higher survival rates. There is certainly a relationship, but it would be idle to claim that an elephant would be unable to rear multiple calves, often, if she were programmed to do so.

But the elephants' high birth rate would be disastrous. The resulting large population would use up all food, starving elephants would wreck the area, and beasts of many species would starve also.

The balanced ecosystem is good, usually providing for maximum utilization of the land and its products by a wide variety of plants and animals, and it is Nature's way. But it is incompatible with the harsher theories about aggression, reproduction, and selfish genes.

Such an ecosystem is often disturbed. Natural disasters, change in climate, habitat or predator, man and his works, may upset the system and lead to a vigorous free-for-all.

Exotic (introduced from somewhere else)

"She could—if she could"

species may have escaped from disease and predators that controlled them in their home land, and/or may prey on organisms that have not developed resistance to them. Rabbits in Australia, gypsy moths and starlings in our East are examples of the first, and chestnut blight and dutch elm disease of the second. Man's travel and trade have greatly increased these invasions. Their impact can be disastrous, but they tend to be eventually absorbed into the system.

Another example of "fit" may usually be found in the succession of plants that take over cleared land, such as an overworked and then abandoned farm field. Typically, the first plants are tough annuals, such as ragweed. In their shelter the perennials such as goldenrod and clover can grow. Then bushes come in, and finally trees, whose seeds are poorly adapted to earlier stages, but will now soon shade out most of the low plants.

There would be great selective advantage to a tree if it could develop a seed that could rival ragweed in growing on poor bare ground. Such a tree might take over such areas first, replacing the succession. Why can't it?

Why are plants such as goldenrod in the cold fringes of temperate zones still frost-sensitive, after growing for thousands or millions of years in such frosty areas? A freeze will often kill them at their height of productivity of flower and seed. Natural selection should favor the development of frost resistance in these, as in thousands of other plants. But they have none. None.

Is there some natural law that limits the total vigor and aggressiveness of a species? Because a tree can shade out competition and live long, must it be deprived of too-efficient seeding? Does the vigor of ragweed prospering on barren soil doom it to annual death?

Why do mammal females become sterile in middle age? Concede that the whole operation of conceiving, producing and caring for young becomes more burdensome with the years, could that matter to either gene or selection if even a few offspring, borne in old age, survived?

These matters still fall in natural selection's twilight zone. Many, perhaps all, are arguable. Now we will get to more serious ones.

The complexity of organisms has always been a problem and a challenge to a simple pattern of evolution. Now, with the development of sophisticated tools for study down toward the molecular level, the known complexity has been increased at least ten, perhaps hundreds or thousands of times.

Consideration of major problems of relying on natural selection begins (and will also end later) with a tiresome one—mathematics. Changes should occur most rapidly with simple creatures with few genes, in which any change is conspicuous and possibly effective, and directions of variation are limited. But the fossil record tells us that evolution was very, very slow when life forms were simple, and is faster and faster as they get more complex.

I have made up an example, starting with a single fact of just-occurred evolution caused by one important trait, and have added a couple of fictitious variations, to demonstrate the resultant slowing and blurring of change depending on natural selection.

The English peppered moth has recently made a change that is a clear example of evolution in action. This insect had either of two wing colors, both speckled—pale grey or almost black, with the grey in the majority. These moths rest on tree trunks, most of which used to be light grey. So the light moths had protective coloration.

Industry's soot darkened the trees, so that the light ones became clearly visible to birds that liked to eat them, and did. The greys became scarce, and the soot-matching ones became a majority.

This was simple, effective evolution by natural selection, which had just one characteristic to select for. But what if a necessary food organism had left the sooty bark surface, and moved down into cracks, where only a long tongue could reach? If half the moths had such tongues, there would be another selective situation, where long tongued blacks would have a clear advantage, but long tongued greys might compete successfully with short tongued blacks. The change from light to dark would be slowed or even stopped.

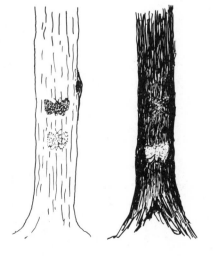

Then, the sooty surface might contain a poison, that would work into the moths through their feet. Again, half the moths might die from the poison, and the other half be resistant enough to survive, with no relationship between resistance, tongue length, or wing color. Now the fittest must be dark, long tongued, and resistant, and the take-over of darks would be further delayed.

If there were a linkage between dark wing and either short tongue or soot weakness, the darks might have stayed in a minority, if linked to both it might have nearly died out.

These are the complications that natural selection would meet with with only three characteristics to deal with. Most organisms have dozens to thousands of survival-related traits. Yet evolution moves rapidly, far more rapidly than when organisms were simple and possible directions of change were few.

There may be a way around this problem. Perhaps the genes themselves have evolved or are evolving so that they mutate more often, and perhaps more purposefully, than in the ancient past.

I cannot offer any suggestion about how it could happen, but genes are such improbable bits, with so much in structure and operation that we do not yet (fully) understand, that it would be risky to be absolutely positive that they couldn't. And if they had, it would remove a very awkward difficulty in evolution theory.

As a further step into never-never land, it seems almost necessary to the successes of evolution that there be some influence, some communication, that has been getting

through to the genes seemingly isolated in sex cells, from the body or the environment. Extremely thorough studies and tests have found no way in which this could be done.

... step into never-never land...

But not-found and no-way do not prove something to be not-possible and non-existent. There still is a faint, very faint, possibility that it is done in some way that has not yet occurred to us.

The accompanying wartime tale relates the achievement of an unrelated but apparently impossible feat of communication, a mystery that is still unsolved.

Through life's early periods, evolution seemed to be so slow that the primitive organism had many generations, perhaps millions of them, during which change (observed in the fossil record) could start, be refined, and spread through a population. In more recent times, it seems often

Case of impossible communication

One night during the last war, after months of preparation, an impressive United States fleet of ships left Adak, Alaska, to conquer the nearby (250 miles) island of Kiska, which had been held by the Japanese for some time.

The fleet, which included samples of everything from torpedo boats to battleships and carriers, operated under a radio blackout. This forbade any radio operator to send out any message to anyone, except possibly from the Admiral's cabin to the President. It required that radio operators stay on duty, nevertheless, to take any incoming messages. Any outgoing message would therefore be automatically detected by other ships.

The Japanese did not wait for a test of strength, but quietly departed before our ships arrived. We had nobody to fight, nothing to do except land and run up a flag. A real letdown.

Radio blackout was maintained for almost a day, while the public relations officers argued about how to break the news.

But we of low ratings who were left behind on Adak knew about the absence of the enemy within an hour of its discovery. The news filtered up to our officers by suppertime.

It seemed certain that no radio or other message could have been sent or was received. But somehow, we knew. Promptly.

doubtful that reproduction has occurred enough times for a particular genetic change, or group or succession of changes, to take over as they have.

Or a change may be so complex that recognized processes could not accomplish it at all, ever.

Skeletal remains of pre-humans (or perhaps proto humans) seem to indicate a growth of brain from near-ape size to near-human size in about two million years, which would represent about 120,000 generations. A prominent biologist-evolutionist wrote recently that this was not a problem; larger brain was somehow needed so it grew. But it is not that simple.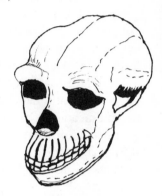

The large and nearly human brain that evolved is considered to be similar to ours. It should have provided many new functions, such as speech (which is extremely complex, and takes a lot of brain space), ability to do mathematical calculations, appreciate music, and many others. Some such functions served little or no purpose then, and might be idle for hundreds of thousands or millions of years until advancing culture made them useful.

The reason and the means by which the brain evolved functions that were purposeless until culture caught up with them has been argued, thoughtfully, even by Darwin himself, with Wallace, and seems to have been set aside with no decision. But at that time, they could not realize the greater problems to come.

Modern techniques have found far more functions in both brain and body than were even guessed at in the last century. But more than functions, the sheer number of our brain cells and cell connections is beyond what our conception of evolution can handle.

Estimates of the number of cells in our brains seem to vary widely, from ten billion up to a hundred billion. If we take the low figure of ten, and assume that the pre-growth brain contained four, there would be an addition of six billion cells during the (guessed-at) two million years. These cells would have connections with other cells; one cell may have from a few hundred to many thousands of these, possibly averaging a thousand. This raises us to six trillion (six million million) connections.

Brain cell connections are extremely intricate and precise. The body had already mastered the technique of making connections; but presumably had little background to decide on new locations; there had been no similar construction before, so it had to be "newly wired."

There was also the problem of providing sources, control, and distribution for some chemicals absolutely required for fine-tuned nerve action. Also arranging for proper space inside the skull, and enlarging female pelvic opening for passage of the enlarged infant skull.

If brain growth were spread evenly over the one hundred twenty thousand generations in two million years, the genes directing the operation would add at least five million connectors and their cells in each generation, to each individual taking part in the advance. Work would always have to be fitted in smoothly with previous structure. Connections intricate beyond our comprehension, far more complex than the most advanced computers, would need to be done precisely. Again, at least five million of them, all at once. Perhaps ten times that many.

Random gene mutations were the only source of change. And life went on, with day to day survival always a priority.

Clearly, natural selection could not have achieved this brain enlargement in two million years, nor ten million, nor in all the time that has ever been.

This is not disturbing to the fact of evolution—our

brains DID evolve—but it means that science should seek sincerely to find the natural laws that make it possible, not continue with the pretence that present theory can explain it.

...nor in all the time that has ever been.

This discussion does not involve belief or evidence that evolution works differently for man than for other animals. It is just taken as an impressive and well documented case, which can doubtless be matched by searching studies of mammals and birds.

One such study might be of the alteration of some land animal or animals into the sea-going seal and whale families. Here again the forces of change had to break new ground: leg to flipper or to flesh-buried remnant, nostril to sealable blowhole, idle tail to powerful driving flukes. To cap the puzzle, the change from birth on land to birth underwater. Think about how that could be done!

Now for another very basic question: what are the selective advantages of the deterioration of old age, and of death? They fit well. Each generation dies to make room for the next, and aging smooths the transition.

But natural selection based on individuals would favor prolonging fertility and life forever, or until cut off by predators, disease, or accident. Why should selfish genes—or any genes—arrange to die?

Our possible ancestors the protozoa, our ancient relatives the bacteria, and our unwelcome visitors the cancer cells, reproduce by division. A complete organism divides, its halves grow, and divide in their turn. Their flesh lives on through endless generations; there is no senility or pre-programmed death.

But every organism that is large enough for us to see and handle, dies. Way back in our evolution we gave up self-division and turned to eggs and sperm for reproduction. At that time, our flesh gave up eternal life—why, we do not know—so we are programmed to grow old and die.

Again, that fits well into the grand scheme, it is good for us and for the world we live in. But how was it done? It does not seem that either selfish genes or natural selection would have made or even permitted that development, if they controlled evolution. Or had much influence.

Lately, much of our thinking should have been disturbed by a curious discovery. It is reported that the natural life span of most species of mammals (but not man) has a close correlation with pulse and respiration rates, which in turn are determined by average body weight.

Now in 1984, still in the early stages of investigation, one researcher reports that during natural lifetime, each non-human mammalian heart will beat about a billion and a half times before it stops naturally—mouse, with rapid beat; medium deer and wolf, and slow beating elephants' (whales have apparently not been checked)—and all lungs expand and contract nearly 330 million times, from birth to natural death. The rates per hour measure out—predict—the length of life in days or years.

There are discrepancies. For example, domestic cats apparently tend to outlive heavier dogs. But the absolute exception is man, who does not fit the scheme at all.

Man's heart and breath timing are similar to those of other animals of his weight, but he will pulse and breathe many, many more times than they will, and so has a life span that is multiples of theirs.

This brings two sobering realizations—that in the midst of should-be randomness and variety there may be important fixed laws that have no apparent reasonable or selective basis, and that man has differences from other animals perhaps more basic than brain, erect posture, and hands.

5. Psychology

Psychology is the discipline (not really science) of the human mind, mental phenomena, and human behavior, and is supposed to include side looks at the same features in animals. It includes or influences psychoanalysis, psychiatry, and education.

This discipline is unfortunate in having very few (or no) clear and un-arguable facts to go on. Its history is theory and argument, counter-theory and counter-argument, on hundreds or thousands of subjects in millions of words.

There are over a thousand journals that are devoted to one or a few of numerous schools of psychology and psychiatry. In many of them, expressions of personal, school-of-thought, and cult origin seem free of restraint. Few or none of them seem to present a general picture of the range of knowledge, activity, and opinion in the whole field.

Any attempt to make an outside assessment is difficult and doomed to failure. In expressing criticism and resentment, the critic must be aware that his message has limited application, and may be something that one of their schools or journals has been saying for years.

I am reminded of a mother giving a bath to her overactive child. She puts him/her in a tub of soapy water, and washes whatever part is in sight. After a few minutes her conscience is almost clear.

Psychologists have done a vast amount of expert research into the minds of humans and animals. Much of it has been detailed and tedious; only occasionally repaid by the flash of insight, the triumph of discovery, that makes such work worth while. But to those interested, the construction of a working pattern of memory, for example, is more absorbing than the best mystery novel.

... as unscientific as Creationism...

However, at the executive or perhaps professorial level, where these and other findings are made up into general theories, there seems to be a compulsion toward bias and error. I believe I am right—or at least not far wrong—in saying that general patterns and conclusions in the main stream of Psychology include the following:

1. Disinterest in human pleasure and happiness, and the paths by which they are achieved.
2. Unstated agreement with the popular idea that increase in possessions, services, and freedoms results automatically in increased happiness.
3. Secondary importance; and a poor second, of inheritance (genetic factors) in shaping human mind, emotion, and personality.
4. Separation of humans and other animals in regard to these three factors is absolute, with any apparent resemblances dismissed as anthropomorphism.
5. Discipline and punishment are ineffective in regulating long-term human behavior.

For the first two, it is difficult to marshal arguments against disinterest, and against drifting with a popular trend. The desire to be happy and the desire to be wealthy are such basic human motivations that our psychologists have apparently not noticed that they may be unrelated, as many of the rich often are less happy than the poor, even the hungry poor.

The main unconsidered (although widely known) factor is that we grow accustomed to what we have. Much or most of the pleasure in a possession is in anticipating and acquiring it, and in its first period of use. It is then likely to gradually sink into a background, where it is only occasionally appreciated, and where it has less and less effect on the normal cycle of mood and activities. In the long period, some personalities will continue to always have some pleasure in each item owned; others will be weighed down increasingly by ownership and care of more and more possessions.

. . . neglect or ignorance

These are deep considerations, important to Psychology in its true meaning, and usually ignored.

Moreover, this type of neglect or ignorance runs deep and far. A few years ago our top scientific organization, the American Association for the Advancement of Science, had as the theme for its annual meeting the topic, "Science and the Quality of Life." According to the titles, none of the reports scheduled for that meeting dealt thoughtfully with any aspect of life quality. They seemed to assume that general advances in conveniences, and in control over Nature, represented improved quality.

In what is probably our finest and most complete assembly of knowledge, the Encyclopedia Britannica, the subject of Psychology has dozens of pages. But happiness rates only a part line, "Happiness: see eudaemonism." Eudaemonism does better, one-seventh of a page. References under it indicate that the last solid word on the subject was about 2300 years ago, when Aristotle wrote that "...happiness is activity in accordance with virtue."

Clearly not satisfying to us. Can't we do better than that — or at least try to? Or even pretend to be interested?

To a certain extent, these are in-house arguments, as to be happy or sad is still largely an internal matter. The primary losses in misunderstandings are to Science itself.

The next two subjects will be dealt with below.

But the fifth, denial of the possible effectiveness of dis-

I was driving among low mountains in southern Spain. I passed a poor farm, with a tumble-down house, and irregular, rocky fields. The farmer was plowing near the road, with one thin horse. And he was singing.

I wondered, "Have I ever heard a union man sing at his work?" and had to answer myself, "No, never."

Then, whose life has the higher quality? The man (woman) who can sing at work, or the preoccupied owner of a car, television, and motor launch?

---

cipline and punishment, is as tragic as it is untrue. Its wide acceptance is a major factor in educational difficulties, delinquency, law enforcement, and many other problems. Unfortunately, there is not enough space here (nor do I have the skill) to take it apart properly, so I will go on to rejection of heredity.

It should be obvious to anyone that human beings, aside from identical twins, differ from each other in many many ways. It may be that each item in our physique is simply a standard item, to be found in thousands or millions of others. There may be a limit to Nature's diversity in size and shape of nose, color and growth of hair, build of elbow, wrist, and fingers, but there is no practical limit to the extent that they can be (and are) mixed up in fashioning a body plan for circumstances to work on.

It is reasonable, even necessary, to assume that unseen parts—brains, brain connections, glands; and organs that influence thoughts, skills, and emotions—are similarly varied. This assumption is backed both by everyday observation, and detailed tests and anatomical studies.

But the fathers of psychology decided that we are all almost identical in the action of any thought-related structures or processes (the blank slate approach), so that any attitude or behavioral differences between people are due to their experiences, largely infantile and childhood ones.

This assumption is almost as unscientific as Creationism, and has just as little right to be called science. Promoting and defending it has probably cost psychology more than half its energy over the years, and a larger share of possible usefulness and public confidence.

But denial of hereditary influence has been promoted and defended with vigor and skill. Those who disagreed were somehow put in a defensive position, and called upon to prove the existence of inherited differences, where it was clearly psychology's burden to prove sameness. And in this area, any hard proof is readily drowned in words.

I was lunching with a graduate psychology student. At a nearby table, an acquaintance shouted unpleasantly at a waitress over some detail of his meal. I commented that his father had a rotten temper too, and was asked, almost nastily, what I was trying to prove.

Not all schools of psychology have gone this far in print, but it has been the firm underlying principle. Why such an absurd approach was chosen is not clear, but it seems likely that they thought it enlarged their potential—heredity offers little chance for change and improvement, except perhaps on an animal farm, while environment's effects on human behavior can be denounced, argued about, and meddled with forever. And it must have made life simpler to dismiss the more difficult part of their problem, by pretending that it wasn't there.

At present, psychological writing may seem to deny or gloss over the bias against heredity, or, as they phrase it, genetic origin. But it is there, and it has spread to take firm hold on associated fields, such as sociology, which may become more fanatic than their teacher.

There is a recent book in which the author proves to his own satisfaction that all mental and emotional differences between the sexes arise from the ancient practice of bringing up little girls as if they were girls. We have only to uproot and destroy this bias, and we will have the exact sameness of the genders that somehow seems desirable to these people.

He seems to dismiss physical differences as having little importance, and anyhow being correctable (yes, correctable!) by early surgery and adjustment of glandular balance.

The sociobiology uproar described next is another indicator of the way that many of these people think and feel.

In 1975, Edward O. Wilson, curator in entomology at the Harvard Museum, produced a magnificent book, "SOCIOBIOLOGY, The New Synthesis." It is a big book, chiefly a detailed and reasonable argument that the forms and activities of complex societies, from ants to humans, are largely determined by genetic factors. It is very convincing, in spite of (my idea) over-emphasis on natural selection.

This book might be properly considered under Evolution. But attacks that were made on it make it a case history in denial of heredity, beliefs of science, and hysterical conduct of scientists.

The author was subjected to widespread and violent criticism, including pickets and physical assault. Most of the attackers (critics) were accredited scientists, and some were on the Harvard faculty.

These developments were surprising. Facts and findings presented by a reputable authority, together with an interesting theory based on them, were condemned with unrestrained fury. This was not on the basis of possible inaccuracies, but only because the material was disturbing to current beliefs and pet theories, which seemed to be thought sacred and untouchable. Resentment exceeded the actual content of the book, and critics seemed to have to even edit and misinterpret the text in order to supply substance to their charges.

A distressing or perhaps amusing aspect is the nature of these charges, which were suitable for temperate comment in a friendly book review, not for the all-out condemnation that was intended. They included:

1. Book justifies present social order
2. Author's position is extremely hereditarian (or deterministic)
3. Author alleges that male dominance is (or may be) an inherited trait
4. Author alleges similar bases in human and insect societies

I hope that the first, a total condemnation of our present social order (they say it shouldn't even be justified), is limited to this particular group, and is not a widely held attitude in the scientific world. Most people think well of both our culture and scientists, and like them to be friends.

The second and third criticisms (condemnations?) justify my bringing SOCIOBIOLOGY's foes into this chapter. How has psychology succeeded in brainwashing other sciences so that they can feel

They did it to Galileo...

inheritance is not only unlikely, but offensive and immoral? Because all of us ARE products of our parents' genes.

The third charge has another feature. It relates to the misdirected drive of feminist groups to make women and men the same; they are beginning to get bored with just equality, and insist on sameness. There are more comments on this in the MICE section. Anyhow, it is clear that any gene labeled FOR MEN ONLY will be automatically resented for this reason also.

The fourth is only a reminder that anthropomorphism is out of style (see long discussions below).

There is the question of how so much sound and fury (there was plenty of it) could be aroused by a learned comparision of insect and human societies. The attackers must harbor real uncertainty about the validity of their own theories, to defend by counterattack before they were involved. "The wicked flee when no man pursueth."

Well, they did it to Galileo, so why not to Wilson?

On the lighter side: My wife, losing in an argument, might cry out, "Don't confuse me with a lot of facts."

Back to psychology's little ways.

Animals are interesting to psychologists chiefly for comparison to humans. There are periods during which they were highly esteemed as being close to humans in various ways. This is called the anthropomorphic (identification of humans with animals) approach. It has been way out of style for more than 30 years, and has zero-zero rating now.

Three articles were printed in a local newspaper in late 1983, on the subjects of human dreams, self-consciousness, and guilt feelings. In each one, as a casual side statement, the article said animals don't have any.

These were extraordinary findings, even for psychologists, and indicate the writer (or his informant or boss) has no pets, and is not well acquainted with people who do.

A sleeping dog acts (sometimes) as if he (she) were dreaming. He goes through a series of actions that cannot be explained in any other way. There is no evidence — anatomical, behavioral, or otherwise — that he is not capable of dreaming. Therefore, it should be assumed that he dreams. But some prejudice, perhaps the writer's, is against dreaming animals, so we are told firmly he can't dream because he isn't human. Unimportant, but an entirely unnecessary piece of false information.

The article on self-consciousness stated clearly and positively that it is a purely human thing, and that animals do not — cannot — experience it. But in two breeds of dogs, spaniels and setters, self-consciousness is one of their most conspicuous traits, and is obvious on even brief acquaintance. Also, I had a large French poodle who was five years old when he received his first fancy haircut, with shaved areas, and a pompom left on his tail tip. The poor beast skulked behind furniture and avoided people for several days.

Then we get to guilt feelings, which again animals can't have. The article says that if they look and act guilty, it is just fear of punishment. In my circles, an adult dog is practically never punished severely enough to bother him; and even if he were, it would take consciousness of guilt to make him fear it. Why else?

In any case, a returning house-holder can usually tell by the first look at a pet if there is a puddle, a pile, or perhaps a chewed chair leg waiting to be discovered.

In each article, the writer made a positive statement that is clearly untrue, misleading any reader who believes him (her); just to uphold psychology's myth of the moment.

This is part of a modern problem, pollution of information with misinformation.

In animal study and comparison, angle of approach may be important.

For example, courtship among apes does not compare with human courtship. Even while conceding that the motivation is the same, the critic finds that the apes' (usually) rather rapid and direct approaches, often including somewhat stereotyped moves and counter moves, do not compare with the frequently delicate, leisurely highly complex courtship behavior of people.

... cripples study...

But looked at another way, it can hardly be denied that the human version is simply a group of elaborations added to, or altered from, the ape pattern, leaving strong and interesting similarities.

There is also a matter of parallel patterns. A thoughtful person who does not allow himself to be discouraged by the wide evolutionary gap between man and other mammals, as perhaps by dogs or squirrels, may find useful knowledge in their coping with human-like problems in human-like ways.

At present, speculation and reports along this line are practically forbidden. No parallels or conclusions must be drawn. Devoted people who spend months or years living with apes or other animals to study their ways must, in reporting, be very careful to avoid human comparisons, or suffer automatic, dogmatic criticism, with probable non-publication of reports.

Such prejudice-and-censorship is not new. In my letter on the facing page, it is shown to have been active twenty years ago, with an appeal to a "basic canon." A canon is an established rule, law, or principle, especially a body of rules enacted by a church council. It conveys an impression of smug rightness not yet justified in any scientific field.

However, prejudice is not entirely without foundation. Many people love animals, particularly their own pet animals, so greatly that they give them much more credit for

**Excerpt from the author's letter
of July 6, 1965, to SCIENCE magazine**

*On 13 March 1964 you published an excellent article
describing research on a fly's nervous system (Microscopic
Brains by V.G. Dethier). This was followed on 26 June by a
letter from William Tavolga, criticising the findings in the
article on the basis that it violated tradition and semantics.
His remarks must be taken seriously, as his attitude seems to
be shared by the majority of workers in natural history.*

*Tavolga cites "a clear violation of Lloyd Morgan's basic
canon on interpretation of animal behavior," and calls
attention to an alleged anthropormorphic trend in the discussion, in much the way that a medieval scholar would
accuse another of heresy. He condemns the use in animal
study of terms and concepts rooted in human psychology.*

*Tavolga's basic stand is that insects and men cannot be com-
pared, as they represent too far a divergence in evolution
and organization. Such differences are extensive, but there
are admitted resemblances in life, protoplasm and environ-
mental forces. In the present state of our knowledge, the
"basic canon" represents dogma rather than a scientific
conclusion.*

*Much more research will be required before we can even
begin to know to what extent differences outweigh resemblances. And such research is greatly hampered by the automatic application of the epithet "anthropomorphism" to any
report of thoughtful investigation. That this word should
have come to be an epithet is in itself unfortunate, for it
surely has a right to be included in the method of multiple
working hypotheses.*

*Our only frame of reference in study of insect brains is our
knowledge of human and other animal brains. If we do not
use this knowledge and the words in which it is recorded, we
will be gravely and unnecessarily hampered in both study
and reporting.*

like-human and even superior-to-human characteristics than they deserve. Separating the core of truth from the fond embroidery of their reports is difficult, so a defensive position of total rejection is taken.

But this policy for keeping things tidy is an unscientific one of throwing out the baby with the bathwater, and cripples study in a very interesting field.

We know of the endless argument, often bitter, about the relative importance of inheritance and culture in determining our behavior. In our present anti-anthropomorphism stance, it does not seem to occur to anyone that pets might shed a bit of light in our dark areas, or perhaps

in places that we did not even realize were dark. Or if the idea did occur, and was diligently researched, it probably could not find a publisher.

Yet if we follow current precepts, each pet dog, cat, parrot, or whatever comes to us with most of its wild-animal ancestry intact in its gene packets, but is given the problem and opportunity of taking its small part in an extremely complex culture.

We hear about the laboratory animals that drool when a bell rings. Stimulus and reaction of pets is more elaborate. The ways in which they adapt, or cannot or refuse to adapt, may give deeper insight into their thinking (and ours) than most lab experiments.

My partner has an old Boxer dog, named Caesar, who has come to work with her for many years. I feed him scraps of various kinds, of which he seems to value most highly empty cream containers and bones from prime ribs. All scraps are given to him in the kitchen, and all except these two specials, he eats there.

But big bones and cream containers are brought into the office, and dealt with at leisure and in luxury on its rug.

Recently, the rug was treated to a long overdue shampoo. In celebration, it was decided that Caesar should henceforth eat all his tidbits in the kitchen. Soon he was presented with a succulent rib bone, and made repeated efforts to take it to the office. When finally convinced that he could not, he laid down and looked at the bone sadly. Even rubbing it on his lips did not inspire him to start work on it. Evidently, a basic principle was involved.

Of course, he won. That, and later treasures, were enjoyed on the rug. But the point is, he has become so over-civilized that his concept of proper circumstances can be stronger than his still vigorous appetite.

Some of us are acquainted with people who have reached late adulthood (old age?) under fairly stable conditions. Among them, it may be noticed that disproportionate emphasis may be put on proper setting and background for activities. For comfortable eating, it may be necessary, or at least highly desirable to them, that circumstances and setting be proper. Perhaps a white tablecloth, certain china and tableware, candles or other soft lighting — any of a hundred different items or combinations that became accustomed and then necessary.

Although we are usually inclined to think kindly and sympathetically of such people and such needs, we cannot help regarding them as symptoms of over-culture. Peo-

ple, we may think, tend to become over-civilized, dilettante, too inclined to stress appearance rather than basics
— a weakness, perhaps small, but definitely a weakness,
and (certainly?) very human.

But here we have a dog, no tradition, no culture, just living with those who have, who is so civilized, perhaps so
dilettante, that he will not (cannot?) tackle a longed-for
bone unless it can be on a proper tablecloth (rug).

I offer no conclusions, except that the refusal to consider animal behavior is among the many grave, almost
inexcusable mistakes, that Psychology makes, and seems to
carry with pride.

6. Right to Life

One of the curious side growths of modern psychology-philosophy is Right-to-Life, an idea that every human, born or unborn, has a fundamental and un-arguable right to live, that overrules the rights and needs of parents and society, and then cancels any need or right to die.

This concept is now widely held to be self-evident, and basic to civilization. Publications as careful and accurate as the New York Times base discussions on it as if it were really a fact of life.

But it is no such thing. It is a recent fabrication, without important roots in either religion or philosophy. The term itself seems to be quite new. Even the stern opposition of the Catholic Church to contraception is a recent (1913) addition to its creed.

It is reasonable to assume that a competent and well behaved person has a right to live. But such a right is not just a gift of heaven or biology — it is built up in the person by himself, as he (she) lives, suffers, strives, is useful to society. It is a lesser right than that of society to protect itself. In no way does it merit the zeal of its promoters.

It is hard to see how such a concept can be extended to cover the newborn or the unborn, and it certainly has no connection with the adventures and misadventures of human eggs and sperm, except in the irrationality that is permitted to religion.

Nevertheless, a right to live commencing before birth is widely used in arguments about abortion and contraception, and has been given status by the Supreme Court, by relating abortion procedures to the number of months from conception.

The more rabid right-to-lifers extend the concept of a living human to the egg and sperm that contraceptives do not permit to get together. They may soon deal with basic contraception — a woman successful in keeping her knees together. The step that seems unavoidable with their logic is to proclaim that any woman who refuses intercourse with any man is murdering his child. She may be on the way home to her husband, but that will be irrelevant. The particular human being who might have been fathered by the man she repulses has been killed — murdered — by her refusal. Right? But at least they are trying to protect murderers — even this one — from execution.

Right-to-life enthusiasts also effectively extend their prejudices into the cruelty of preserving children who are seriously defective at birth. The children, the parents, and the community would all benefit by their elimination. Even when ways might be found for their repair, they will usually be risky, and fearfully costly in money and in pain.

The world could probably support almost a half billion people comfortably, after allowing space and cleanness for Nature to maintain her other works. But we have over four billion, with increases every day. With this condition existing, right-to-life people are on the wrong side, they are evil — they oppose human and world welfare with an empty and hurtful concept, which they seemingly have created in the last few years.

**Murderess of
the unconceived**

How can it be said seriously that each of the millions of babies born each day — or maybe only each week — is a sacred and irreplaceable chalice of God's will, with an absolute right to live? We are the product of a long evolution — millions of years long — during which crippled or even surplus newborns were automatically killed or allowed to die.

Apparently for the purpose of making the killing of the newborn less difficult for the mother, Nature has provided a lapse of about a day between birth and the affection-stimulating act of nursing. Christian baptism has always been at least a day after birth, again perhaps to allow time for uncertainty as to whether the infant lives or should die.

... *cruelty of preserving children*

We should respect and keep the wholesome wisdom of care and sacrifice that produced us, however repellent it may be to the artificial standards of pressure groups.

Money should — must — be considered. It is the custom of those seeking to expensively rebuild, treat, and maintain defective babies to speak with withering contempt of those sordid characters who would even think about money when a baby's life and health are at stake.

But the investment, which may be huge, seldom does anything for the babies, except to cause and prolong suffering. At best (or worst) it may keep them alive, perhaps crippled, stupid, in life-long pain. If they could think clearly, they would long to leave, every endless day. For this, should society spend hundreds of thousands or millions of dollars on individuals, and tens of billions to try to do them all? A society already struggling with more debts than it can ever pay?

Nor do we have the people. How cruel and wasteful to ask skilled surgeons, trained technicians and devoted nurses to waste their lives this way, when conscious people need them!

There are parallel situations made by the same fetish for preserving life, however grim that life may be.

I receive mailed appeals from people trying to feed starving children in a southern Sahara—Western Sudan region (Sahel?) in Africa. The pleas and the pictures are heart-rending, and the money needs per child are small.

But children that are fed grow stronger and sexier than those who are hungry. As a result of the extra food we may supply, there will be two to four times as many new children within the next twenty years. The land is poor and dry, already over-worked for crops and pasture, and cannot provide any important increase in food. So those children will starve also.

So those children will starve also.

It is a bad thing that a child should be hungry. But is it not worse that three should be hungry? The result of devoted work in Africa and flow of sympathy money from contributors is bad — very bad — multiplying a problem instead of solving or even reducing it.

Barring war or pestilence we are surely increasing — greatly increasing, three or four to one — the misery that we strive to reduce.

It is shocking to realize that the work of these good people is good only if human life in the world — or perhaps only in their part of it — is to be destroyed within a few years. If so, present pain is relieved, without the future cost of tripled pain. But that solution is even more intolerable than the problem itself.

This situation is serious now in many lands, desperate in some, and threatening to the whole world. It was predicted around 1800 by a British economist, Thomas Malthus, who pointed out that population increased by multiplying, food supplies only by the slower process of adding, if at all.

We have kept ahead of his warning all these years, by better and better-yielding crops, by irrigating deserts and destroying jungles. But possible crop land is running short and is deteriorating; we are approaching the ultimate in crop yields; and human multiplication continues on its remorseless way.

There is a smaller, non-human reminder of this problem on our doorsteps.

We have killed most of the predators — wolves, panthers and others — that used to kill and eat deer, limiting their numbers. So now the deer's reproduction rate, set up to absorb vast losses to these animals, is too great. They multiply, in most areas of the United States, until they do not have enough food, particularly in winter.

A starving deer herd is misery, whatever its scale. Hundreds or tens of thousands of beautiful animals; thin, cold and hungry, sometimes just standing pathetically, sometimes grubbing roots of next year's food bushes, occasionally nibbling gifts of hay that they cannot digest, stripping trees of bark, roaming into settlements to eat them bare of prized plants; and getting weaker, dying by the dozens or the thousands.

But in hunting season, a few months before, earnest groups, mostly women, may have prevented hunting these same deer — by argument, picketing, blocking roads, getting hunting licenses themselves to prevent their use — although they must have known (at least they were told, repeatedly) that the animals they kept from a quick death and human use would probably, as a result of their actions, starve to death, miserably. How can they? It is the same mindless drive that must save Baby Doe, regardless of what she and society must suffer as a result.

Perhaps someday they will realize how cruel they are. In the meantime, the unnecessary suffering goes on.

At the other end of life we have the old, often condemned to seemingly endless years, perhaps full of pain and confusion, and often bitter knowledge of uselessness. Here almost everybody, not only the right-to-lifers, is determined to keep them alive.

I saw a headline on a magazine article, "How to Help a Friend Who Wants to Kill Himself." I thought that here at last there might be something kind and helpful, perhaps a diagram for wrist cutting, or just a list of good over-the-counter poisons. But no, it said to sit on his head and call a cop.

... *the mindless drive that must save Baby Doe*

Nowhere in that article, or any other on the subject that I have read, did it seem to even occur to the author that a person might have a right or a reason to kill himself, nor that it could be a good idea.

But the anti-suicide people are not just a nasty little pressure group — they seem to be almost everybody. So I give up, and turn to another subject almost as hopeless.

Capital punishment — killing a criminal as punishment for or in expiation of his crime — is a natural target for criticism from right-to-life groups, and may be even more vigorously opposed by many others.

Such people often say, loudly and falsely, that society has

no right to take the lives of any of its members. But most societies, including many religions, have, from the beginning of history, killed their members quite freely. Causes for execution have included murder and lesser and even minor crimes; treason, heresy, religious ceremony.

It is likely that a majority of our Right-to-Lifers, and opponents of capital punishment, have either Christian or Jewish faith. They should respect the Bible's Old Testament, and should read the Book of Joshua.

In it, the conquering Hebrews, with God actively with them, captured many cities. In each of them, as a matter of routine, they killed every man, women, and child they found, often all the animals, and set fire to the buildings. Battle horses were crippled (hamstrung) and left to die.

They were reasonable by their stern needs. In this poor land, there was not enough food for both victors and vanquished, so the vanquished had to die — not for any wrongdoing, but because they were losers in battle — battle that was hopeless, with God on the side of the invaders and aggressors.

With such a precedent, how can we denounce capital punishment, and refuse to allow the miserable to die?

Hanging is a solid part of our English background, and seems to have disappeared because of a general softening in our emotional reactions, rather than faults in the procedure itself, or its usefulness.

For many years, states continuing capital punishment have sought increasingly private and painless ways to kill. Electric shock took the place of hanging, and now has competition from gas and injection.

Since execution now seems to be entirely for murderers, this preoccupation with privacy and painlessness is absurd. It is a safe guess that the murderer was seldom that considerate of the victim(s).

Public arguments for the last 50 or 100 years have been wrongly weighted. Opponents call on those favoring exe-

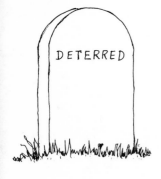

cutions to justify their position, while they themselves are the ones working to undermine and destroy an ancient protection for society, without solid reasons.

One of the opponents' major arguments is that it has not been proved that executions are a crime deterrent. The criminal who is executed is certainly deterred. Beyond that, opposition has seen to it that executions are too rare to have any effect.

In a recent 10-year period, about 2,000 death sentences were imposed but only seven performed. A death cell inhabitant had a better life expectancy than an outsider who smoked two packs a day.

As a special twist, anyone who does work up a good deterrent case is accused of wickedly wanting to extinguish human lives just to impress people.

The impact of executions will continue to be slight because of the endless court delays originating in sentiment rather than justice. Reasonably, the attitude should be casual. Lots more people die than go to prison.

Secrecy in execution of a criminal weakens its effect, and it is hard to justify it. Full television coverage through all interested stations would be much better. Millions of television viewers like (and obtain) violence on the screen, and it is mostly anti-law and anti-society. Hangings would raise the moral tone, and probably the deterrence factor. They should be much better, much more deterrent on camera, than other methods of execution.

The next most frequent argument against killing criminals is that the wrong person might be executed, an error that could not be rectified. True, but consider.

If an innocent man spends years in prison and is then released with apologies, can anything make up for his suffering? Usually not. But an executed man is clean out of it.

Many lawful drivers, proceeding at proper speed on the right side of the road, are killed in automobile accidents. Is that sufficient reason to stop using automobiles?

He was right, dead right, as he drove along,
But he is just as dead as if he were wrong.

According to everything that we surely know, death is a problem only to the survivors. In many, perhaps most, religions, drastic punishment in this life, particularly execution, is a priceless asset to be carried into the next, as atonement already accomplished for sin. A whole life's sins, not only the perhaps uncommitted crime in question.

7. Aging

Mental and physical deterioration is characteristic of most aging mammals, including humans, in whom it has of course been studied in the most detail and with the greatest amount of argument. On the mental side, the process is not understood. (In scientific terminology, "not well understood.") Until recently, it has been mostly called just aging, or senility, and extreme result senile dementia.

One of the puzzles is the reason for aging. This is little discussed because possible answers tend to be disturbing to the firmly established theory of evolution by natural selection. More about that elsewhere.

On the matter of how it is done, we are a bit smarter. We find that our cells are often endowed with ability to reproduce themselves, but under very strict number-of-times limits associated with their type, job, location, and age. They also may die or stop working for various reasons. Surviving cells may produce new cells to take the place of those lost, and replace connections, sometimes accurately, sometimes not.

As we age, the rate of cell loss increases, and the rate and quality of replacement diminishes. At some time, these two rate curves cross, and we start to go downhill in number of cells, and probably in quality of connections also. I have heard it said (by a 22-year old) that this happens when we are 22, but it seems more likely to be in the forties or fifties. And, I am sure, not at one pre-arranged age for everyone. Also, the speed of decline is almost sure to vary greatly among the different functions of any one individual.

For a long time, perhaps even a long lifetime, decline in number of nerve cells may be more than compensated by experience providing increasing skill in using what is left. This seems to be the good fortune of some of us.

Number of cells and connections are only part of the story. Nerve cells, in the brain and elsewhere, depend for action on minute electric charges, and on a number of chemicals that strengthen or weaken response.

... has not wept since childhood...

In a general way, it seems safe to say that the mental-emotional side of senility is partly due to reduced number of brain and nerve cells, confused wiring in replacing those that have died or signed off, and irregular and/or deficient production, distribution and use of chemicals needed for nerve function.

Such losses show up in confusion, short, mixed-up memory, forgetting of well known names, faces, and important experiences, uncertainty as to dates, places, directions, and intentions. There are often day-to-day and longer period changes in competence for better or worse. These things appear in various combinations, readily recognized by those who associate with the elderly.

A man who has not wept since childhood, or a woman who has done it seldom, may weep often in old age, for half-remembered griefs, or the sad condition of the world, but more often simply because something is very beautiful, or someone is unexpectedly kind.

There is tunnel vision — a driver may find it difficult to transfer attention from the road to surroundings (or even to the dashboard instruments) and back to the driving. A result is to go unnoticing past friends, or roadside changes, or even a planned turnoff.

The old may often look directly at something that is important or immediately needed, such as a tool, a bank statement, a shopping list, or glasses, without seeing it, and then spend time, a lot of time, looking for it hopelessly.

Failure of sense of direction may be only in not knowing, or (much worse) a twist. Entering a subway car with seats facing inward, there may be an intention to sit on the left and a last-second change to sitting on the right, that does not register in the mind's compass. Then everything is 180 degrees wrong, eyes raised to look out the windows find the train going frighteningly the wrong way, and the sternest mental exercise (or going up onto the street) may be needed to control the error. Eventually, inattention to changes of direction while driving or walking may result in similar directional confusion.

These are only random samples of the many ways in which aging may affect an individual. When they are severe, and there are enough of them to affect competence, they may be called senility.

Mind deterioration caused by aging usually becomes noticeable to its victims in the sixties, but it may have given indications twenty years before that, with very-difficult-to-measure dulling or straying of thought processes, difficulty with or loss of parts of specific faculties, such as perhaps mathematical ability. Physical deterioration characteristic of the age groups also occurs.

Some people adjust to aging smoothly, reducing mental and physical activity as required, without seeming to give it much thought. Others resent it and fight it. An important factor in their adjustment may be whether self image declines at at least the same rate as competence. If it does not, an individual may greet each new failing or blunder with embarrassment, humiliation, or exaggerated concern, none of which reactions are justified.

An aging person may be told by his doctor and friends that the best way to reduce and postpone the attacks of old age is to keep busy and interested — stay in business or start a new (small) one, become a nuisance in local politics, do a thankless no pay job as secretary for the local golf club, take kids for nature walks, or whatever.

They are right, but there will probably be penalties. If he (or she) follows this advice, his activity will often force him to distinguish between creditors and customers, remember which side of a political argument he has favored, prepare an annual report from papers that keep disappearing, or make a poor guess whether a beastie is a toad or a frog (after majoring in zoology in college).

He must have or grow indifference to these problems (the best way is to realize that they are funny, and enjoy them), or he would be better off in a rocking chair, or in some intermediate status.

There are many good people who feel that simple and direct words about setbacks and misfortunes are likely to depress or offend, so that they choose words or phrases with less direct meaning and often more syllables. For example, it seemed that to say that someone was in jail was harsh, so the word prison was substituted. With increasing sensitivity, that became too definite also, so that now a convict may be merely said to be away.

It is doubtful whether such word changes make much difference, but the intent is at least kindly.

Dear Charles is still away

In this manner, old people became elderly or senile, and then senior citizens. But, rather recently the tide has turned against them. Wide publicity on research into an affliction called Alzheimer's Disease seems to go out of its way to insult, depress and worry the old folks.

This ailment involves accelerated mental deterioration, caused or accompanied by formation of tangles and plaques in the frontal and temporal lobes of the brain. Symptoms are similar to those of ordinary senility, but develop faster and are more pronounced. It is typically a disease of the aged, apparently affecting 6 or 7 percent of folks over 65, but it may also develop in younger people. Its cause is not known, but a slight tendency to group in families indicates a possible genetic factor.

There is no treatment at present, except to try to keep the victim comfortable, which (alas) usually means a nursing home after the disease is well advanced.

Doctors studying this disease characterize any effect, even if slight, that it has on mental performance as dementia and seemingly enjoy extending the term to almost any senile lapse of anybody in memory or in thinking.

They have developed a theory that there is no such thing as senility, even calling it a myth. They claim that in the absence of a definite and mind-crippling disease we retain most of our mental powers, possibly slowed down a little, until death in our eighties or nineties. They say that "so-called" senility afflicts only about 15 percent of the aged. In this fraction, Alzheimer's disease is the cause for about half, and specific ailments such as cardiovascular disease the rest. No natural decline is acknowledged.

The nature of the tests by which they claim to have demonstrated that senility is a myth never seem to be stated. They may very well be inappropriate and superficial, are probably biased, and are certainly inaccurate in following senile thought processes. It seems almost sure that none of these researchers have intimate, relaxed, and unbiased

A Quaker lady said to a friend, "All the world is queer save thee and me, and sometimes I think that thee is a little odd."

outside-the-laboratory familiarity with ordinary old folks, with their quirks, oddities, and special problems that may make them so lovable, pitiful, and/or infuriating.

The no-senility people do not seem really sure themselves. In one article denying the existence of such a state, the author recommends fighting it off by keeping busy and alert. He also tries to be encouraging by saying that just loss of memory doesn't mean that senility is coming. Here he overlooks the fact that forgetfulness may mean that it has already come, as loss of memory itself is its most conspicuous feature, often linked with confusion. If severe, it creates discouraging problems in almost every human activity—businessmen work, social, and even self-care.

For example, a simple operation like taking a shower can become a problem if the showeree can't remember what he (she) has washed, or which way to turn the handles, making very hot water dangerous.

As to dementia, to most people it simply means insanity. It is the original official name of schizophrenia. Dictionaries define dementia as madness or insanity, and a big one has space to drive home the point in these words: "Unsoundness of mind to the degree of total loss or serious impairment of the faculty of coherent thought; stuperous insanity." The word, "Dement" means to impair or destroy the mental powers of; make insane or idiotic."

... no-senility theory is actively promoted

Strong words to describe the old fellow whose principal delusion is that his daughter is still alive. Wouldn't "a bit senile" be kinder, and a lot more accurate?

This dementia term is obviously too strong to be applicable to ordinary senile lapses and confusion, and it is hard to understand why it was chosen by the supposedly knowledgeable and devoted workers in this field. They should have given thought to its side effects, whose severity may be limited only by its not being noticed.

Aging may be a trying experience for many. Other periods of strain and frustration during growing up from a baby into adulthood can be relieved by knowledge or hope of progress, by pleasures of active life, and by striving to shape a future. But getting old is mostly retreat (although with occasional or even numerous rallies) and the end is unarguably death. Comfort may be found in acceptance; pleasure in companionship (with man or bird or beast), in hold-out rivalries with other oldsters, satisfaction of "doing fine for my age," and random activities with decreasing mental and physical requirements.

It can be hard to take, and its victims should be spared unnecessary distress.

Nevertheless, this no-senility theory is actively promoted, in the press and on radio and television, not only by newscasters but by doctors themselves. It is almost certainly false, and even if true, can cause grave harm wherever it is believed, or even taken seriously.

Consider samples of possible broadcast listeners to a no-senility, just-Alzheimer's talk. A man, 60 years old, aware of an increasing number of mistakes in his work, worries about being fired before he is 65, or even 62. A 70-year old can't balance his checkbook any more; another, a bit older, gets lost in a familiar neighborhood park; and his wife has just poured hot water into the sugar bowl instead of into her teacup.

They are each wondering, perhaps only a little, about themselves, but taking comfort in the naturalness of failings occurring at their ages.

What must be the effect on them when they are told by a doctor, speaking on a trusted broadcast station, that there is no senility, 85 per cent of oldsters stay competent until they die, and anyone who is slipping is demented (insane) and has a rapidly progressive and incurable disease?

The chances that any of them have brain-affecting disease is small, less than one in six (on the basis of symptoms mentioned, far less), so they are being unnecessarily subjected to a cause of worry. If any of them have Alzheimer's, nothing can be done about it, and it would be kinder to leave them in ignorance until definite symptoms are noticed by associates or doctor. So why broadcast?

Aging people have a special place in our society. Most of the non-aged make allowances (although they may need reminding) for probable physical and/or mental weakness. Old folks are seldom expected to do physical labor, or to function as briskly and understandingly as younger ones. Inertia (laziness), mistakes, and oddities of dress and behavior are almost automatically forgiven, often with friendly warmth and offers of assistance.

Of course, this has its bad side. Senior citizens who are still competent may be brushed aside and humored, instead of receiving the respect and action that their ideas deserve. This can be frustrating and even enraging.

However, since a majority of very old folks are likely to be somewhat substandard in body and/or mind, advantages to them of kindness and consideration greatly outweigh the occasional hurts. And disbelief and snubbing can happen at any age, particularly during childhood.

Long before I thought I deserved it, I received an ornate birthday card saying on the front, "I hear you are still chasing the girls" and inside, "But do you remember why?"

The next year was milder, only "Don't let them tell you that you are old!" and then, "Hit them with your cane!"

If the there-ain't-no-senility school should convince the general public, this type of tolerant warmth might be reduced. The aged should then be expected to be as bright, accurate and useful as anyone, so would be expected to understand, help, and work beyond their fading abilities. This would lead to over-striving and frequent humiliation of the oldsters, and disappointment of their associates.

Thus, success in such anti-senility propaganda can lead not only to worry and self-doubt in the old, but also to at least partial loss of their rightfully privileged or at least indulged place in society.

85

This problem does not arise from interest in, and study of, Alzheimer's Disease, but in the almost unrelated contention that senility is not also commonly caused by natural aging. The situation would be much better if these Alzheimer's enthusiasts would also undertake accurate and sympathetic studies of natural senility (which has fascinating aspects), rather than waste energy and spread unhappiness trying to prove that it is a myth, which it certainly is not.

...Alzheimer's Disease...

Another drive against serenity in old age is in regard to sex. Some men are already losing interest/ability in their fifties, giving up is quite usual in the sixties, and very, very few keep on to their nineties. Loss is usually accepted more or less placidly (or perhaps even with relief) as a normal part of growing old.

But now we have smart lecturers around us, in print and on the air, saying that practically all men are potent at least through their eighties. This is untrue and anyhow is unlikely to be useful. The men who still can manage usually at least suspect that they can, and take a little pride in the thought that they are exceptional. So this propaganda diminishes their self esteem. Those who can't are depressed by hearing that lack is neither usual or natural.

This physical problem is for men only. Aging women can usually take care of the "ability" with a bit of lubrication, and extent of interest can remain their private affair.

Many old and getting-old people are involved in a problem of retirement from work. As age advances, they may find themselves weary — perhaps tired of the work itself, or of commuting; or annoyed with boss, co-workers, new systems — or perhaps just tired.

The employer might also be tired of the worker. He might have become slow on the job, make too many mistakes, talk too much, or be difficult to break in on a new machine or new system.

A concerned employer may handle either or both situations by a refresher program. This could involve work bonuses, discussions of job importance, job changes, or other procedures. The odds against success are substantial, costs increase, and worker pride and self-respect are likely to be hurt.

The best answer (although far from perfect) is an arranged-in-advance automatic retirement at a specific age. It cannot cure the hurt of getting old nor continue the pay checks. But it avoids the worry-in-advance and the humiliation of being personally fired, and is an honorable retreat from work that may have become too demanding. It may have kept the job going for years of sub-standard work.

Such a system can and should be set up to accommodate very special cases, in which a still capable worker wants to stay on, or a tired one wants to quit early.

With retirement time arranged in advance, earlier discharge (except for depression layoffs) is most unlikely, and years of worry may be reduced or eliminated. Time is afforded for such arrangements as are possible.

Advantages of automatic retirement at an appropriate age, particularly to match Social Security, are obvious. It is hard to understand the people who think they are on the workers' side, yet bitterly oppose it, and even make it illegal on the ground of age discrimination.

8.   Fat

Many people weigh too much. Scientifically, they are said to be obese, and their problem is obesity. In ordinary speech, they are fat, plump, or heavy.

The excess ranges from a few pounds, unimportant unless trying to wear last year's dress or trousers, up to carrying a hundred or more pounds extra.

Fat affects millions of humans, particularly civilized ones, and a few animals. It is the subject of vast amounts of public discussion, and inspires too many people to tell others what they should (or more often, should not) do. It has been generally blamed on overeating, and has spawned innumerable diets.

In some cases, it really is due to overeating. Sometimes not only stuffing down too much food, but specializing in large quantities of fat-forming foods; fats and carbohydrates in any form available. The unfortunates who act this way are usually overweight in a big way.

A large proportion, probably the majority, of overweight folk do not overeat. They may have a special fondness for fattening foods, but may also have the will power to mostly resist them. They can usually lose a few pounds by strict dieting, but it tends not to stay lost. This brings on a lot of shifting among diets.

There are some people, a small and fortunate minority, who have an opposite problem. On visits to the book bindery, I often took the vice-president's secretary out to lunch, with or without her boss. It was fun to watch her eat. She ordered the most fattening foods on the menu, and ate them with zest. I thought she was attractively thin, but her husband thought that she was too skinny, and complained. By excessive eating, which she enjoyed, she could put on a few temporary pounds for him, and the chance to do so at a customer's expense made her very happy.

During the past few years I have read a number of magazine articles about obesity (fat), first with annoyance, finally with stunned disbelief.

In each of them, weight gain or loss in normal people is determined by food intake, together with the efficiency of metabolism in converting it to energy or heat. Any non-consumed food substances are stored as fat. There has been thoughtful consideration of personal peculiarities and rare diseases or structures that affect the rate of conversion.

There is no hint that much of our food does not need to be metabolized or burned because it is delivered to the toilet for disposal. No mention whatever! Can you believe it? Were even the Victorians that prudish?

Excess fat (obesity) is a very real problem to millions of people, and deserves extensive and careful study.

As in too many scientific areas, there are distinct styles and biases, varying from decade to decade, in studying and reporting. Some years ago this problem had a simple approach — overeating is the (only) cause of overweight.

This has changed. It is now widely recognized that most fat people eat no more than their lean friends.

But still, for most the only practical remedy is to eat less. It still may be many years before we understand the problem, perhaps many more before we can cope with it. In the meantime, undereating is the simplest and often most effective remedy for fat.

But dieting MUST be done with moderation and discretion. Lost fat may be accompanied by haggard appearance; disposition and/or health may suffer, perhaps disastrously; both without the overall shrinkage desired.

There are various oddities in human structure and function that can affect metabolism and burn-fat-to-keep-warm, but not enough to provide a general explanation.

As I have complained, the digestive system is not considered. Yet it is well known (or was in the dim past of my college years) that the body can be selective as to both *... nothing is this simple.* kind and quantity of different food substances it absorbs from stomach and intestines; and, within certain limits, it is able to break down its own substance, and excrete the debris.

It is fashionable now, among the experts, to almost avoid the eating part of the problem, and concentrate on metabolism, or rather, on that part of it that is concerned with changing food into energy.

Their formula is beautifully simple. From the food intake they subtract the amount that the body cells consume in providing energy, figured on measurement of the basal metabolism rate (BMR), plus an allowance for consumption to keep us warm (thermogenesis). Whatever is not consumed becomes fat and is stored in the body.

In real life, almost nothing is this simple.

To repeat — because it is important — no allowance ever seems to be made for the part of the food intake that goes through the digestive system to be deposited in the toilet. If they said, "net intake," or "intake less elimination," it would allow for this pass-through. But they do not.

There is body repair. Tired cells, of most kinds, are routinely scrapped, disassembled, and delivered to the intestines via the blood stream, according to various maintenance patterns. Such cells are usually replaced, are usually protein, and the replacement material must be obtained from the net food intake.

But obesity specialists, at least as indicated by their articles and reports, are indifferent to this, speak only of energy-heat-metabolism (as if it were the only kind of metabolism), and give the incorrect impression that fat is the only body structure material being built.

Having thus narrowed the subject, they pursue various oddities. There is an enzyme, ATPase, that works at pumping sodium and potassium in and out of our cells, a humble job but essential. The fat people who were studied tended to have a shortage of this enzyme.

An outsider would think that this would cause cells to be sick, or at least lazy. If there is such an effect it has not been noticed, or anyhow reported. But less pumping means less energy consumed, less food converted into (pumping) energy, and more food forced to store itself as fat.

Another report, which I cannot understand, blames brown fat, apparently a very stable body ingredient of unstated use. Its quantity, or its activities, may influence formation of our ordinary white fat, which is our worry.

These findings, and others, are interesting, and possibly important. But they represent rather uncommon situations, and excess fat is very common.

The control system has not been found.

The important and well-recognized spur to metabolism is activity, but it too has had weak promotion lately. It takes energy to move, so the more we move — that is, exercise physically in either work or play — the more calories we consume or burn instead of passing them on to our fat accumulation.

The exercise remedy, or perhaps counter-offensive, is fairly effective with some people, less so with others, in reducing fat. Its side effects are usually more pronounced and useful than its fat reduction — improved muscle and emotional tone, vigor, and physical skills.

The term metabolism should include two subheadings. The first, catabolism, is destructive. It is breakdown and removal of existing body cells. The breakdown process may provide substantial energy. It is usually not included in the "simple formula" as it is independent of food intake. If overdone it can result in marked loss of weight.

Metabolism's second department is anabolism, which takes simple substances from digestion and builds them into complex body cells. References seem to cite building only protoplasm, but installing fat should be in the same body department.

These two processes are fairly well balanced in the normal healthy human. If the building side (anabolism) is stronger, weight increases; if weaker, it should decrease. The control system has not been found. When it is, it will probably be found to be more important than some (or all) of the eccentricities just discussed. Later for that.

Metabolism is doubtless sometimes very important in weight problems, but it is basically a side issue. Our digestive systems, no longer mentioned by authorities, are the key to the working (as opposed to executive) part of the weight problem.

... basically a side issue.

Our digestive equipment is a marvel of design, although not of beauty. It is a single line disassembly plant, reducing a wide variety of bite-size foods, and liquids, to simple compounds in tiny sizes that blood can remove for absorption into the body system.

The operation is intricate.

The processing of food from swallowing on down is basically automatic, with various sections timed to do their part of the work, then push the material on to the next. However, the presence of strong regulating (or interfering?) nerves is noticeable when fright or even acute worry stop or cripple digestive action. They may have influence always. The body (blood stream) absorbs a little nourishment, and many drugs, directly from the stomach, but most of the nutriment is picked up from the duodenum and the other two sections of the small intestine. These organs produce an impressive number of digestive enzymes, and have means of conveying digested (broken down) food through their mucous membrane lining into the blood.

Digestive juices are both general and specific. Hydrochloric acid, produced in the stomach, attacks most food substances. Saliva, from glands in the mouth and also below, changes starch to sugar. Bile works on fat, pepsin on protein, almost any kind.

In addition, there are specific enzymes for many or even most food substances. Lack of an enzyme is likely to prevent digestion of some specific food. For example, lactase is needed to digest milk sugar (lactose). Many adults of central African descent do not produce it. They should not drink milk. It makes them sick.

Digestion is influenced or made possible by many such enzymes. If all of them are there, things go smoothly. If one is lacking, its particular substance will not be digested. The result might be distress, as without lactase; or it may be without symptoms, perhaps leading to an unhealthy lack in the body of a substance that is in the diet.

... need a top-level search for a master control...

A considerable bulk of food passes through the digestive system unused. Much of the waste may be digestible, useable material. Reasons for its rejection are often not known, and if reports in scientific magazines are accurate indicators, there seems to be little interest in the matter.

Yet digestion would seem a far more productive field of study than metabolism, in trying to find the causes of body weight problems. But in both fields, the methods and aims of current research may be concerned too much with how the detail work is done, not with the real question, WHY?

We need a top level search for a master control system, some bit of brain that manipulates our appetites, and the output and destruction of substances that regulate digestion and retention of our food.

This regulator may operate according to standards determined before civilization, but seems as if it can be greatly modified by our individual inheritance and circumstances, and a variable amount by diet or non-diet; fast metabolism or slow; activity or sloth.

We can see some of the pattern if we avoid too much attention to individuals, and think in terms of masses or averages, usual and unusual conditions. When young, in our early working years, most of us carry little fat. This lean figure may be preserved through youthful years of unemployed idleness, with a low activity level.

As middle age approaches, fat begins to accumulate, usually slowly, sometimes faster. We may get heavy because we slow down, or we may slow down ___ *... here is the message.* because we get heavier. Or perhaps there is no relationship, except in ages of occurrence.

Extra pounds may be one of Natures's ways of reducing our activity rate, to tell us it is time to lay down the spear and do something slower, like tanning hides. In trying to judge, it may be worthwhile to realize that our middle age may have been old age, once upon a time.

The picture is blurred after reaching Social Security age. Many of us thin down, anything from a little to too much. Others never give up plumpness, which may have become desirable as a wrinkle-filler.

There are many, many people to whom this schedule does not apply. Their weight might still be rigorously controlled, but by a defective or perverse mechanism.

In the very fat person who eats enormously, it seems to be stuck at "add weight." It can stay in unusual positions, as in my lady friend at the bindery who could hardly maintain enough padding for her husband. And there are those with fat bottoms and lean tops, or (more rarely) the other way around, in whom it is confused.

Control may be variable and tricky, and enjoy teasing, as by temporary response to new diets. Or sulky—adding pounds because of eating a one-eighth ounce piece of candy. But the general trend from just-past-youth to almost-old is to add weight.

After all the talk, here is the message. Our scientists interested in obesity seem at present to relate its problems only to energy-heat metabolism errors, which are probably less important than the whims of digestion. And the workings of both of these seem subject to an overriding regulation center; the finding, study, and perhaps control of which are vastly more important than the details of how its influence is shown.

9. Mice

A profoundly significant experiment was reported a few years ago. It got space in SCIENCE magazine, but has seemed to attract no interest.

A researcher had observed, as many had before, that mice of the same litter had little or no sex interest in each other when mature, and when mating was induced, it turned out badly.

He mixed up a number of newborn litters, so that the baby mice in each litter were not related to each other. The effects of nesting together on adult sex behavior were still the same. It is evidently growing up in close association, not genetic relationship that, for them, activates Nature's controls against incest and inbreeding.

Mice and people are of course very different. However, they have enough traits in common so that animal findings can inspire research for human similarities.

We have particular reasons to look into this one, as our children are being forced into increasingly close mixed-sex association. Women's organizations, starting with a basically reasonable campaign to obtain equal rights and treatment for women, have wandered into a number of side paths like this one, whose value is by no means clear.

Sex segregation in schools is an old but usually not very positive tradition which had been crumbling long before NOW (National Organization of Women). As of the time of their impact, co-education (boys and girls in the same classrooms) was practically universal in public schools and colleges. Private institutions were frequently one-sex, although there was often a closely associated college for the other sex, with variable fraternization.

Dormitory living was always (I think) sternly one-sex.

Lower grades usually had two playgrounds (in the Northeast at least), one for boys and one for girls. If there was a difference in quality, girls got the poor one, as they played less and cared less.

In school-run activities, the two sexes almost never played together, and seldom against each other.

...possible effect of adult-mingling...

On the often slender ground of difference in playground quality, the NOW reformers in many places are managing to destroy the separate-playground system, to mix the sexes on one or both of them. In this work they seem to be inspired by a dim but fanatic notion (far different from what they say) that success for girls (and presumably for women) can be found only by intimate mingling with boys and men, doing the same things that they do for the same rewards.

If I were a woman, I would be humiliated to tell a school board meeting that my daughter could play rewardingly at school only on a playground where there were boys. That she and her friends would be deprived if her class, her school, or her playground, was all-girl.

But enough of that for now. Back toward the mice.

Throughout the world, in the cultures we have been able to study, we find a strong tendency to separate the sexes in childhood, up to mating age or to mating itself. There have generally been obvious advantages in the various systems, but the mouse litters add a basic one that should not be ignored. Old customs are often wise.

There is a local trait along the same line. Opposite-sex children growing up in close proximity, as for example next door neighbors, usually don't think much of each other, and seldom pair off. This is only my observation and personal experience, but it seems a common happening. Damage occurs up to and through dating age, although the only report on the subject that I have seen limited it to ages zero to seven, for reasons that were not clear.

In considering dangers from too close and widespread association in childhood, the possible effect of adult-mingling should also be appraised. Women normally have time free of men, in clubs and in child rearing.

Men, however, have been cut way back. All-male associations of almost every kind, including posh clubs, men-only plane flights, and miners' work gangs, have been vigorously attacked and infiltrated on at least a token basis.

My nephew Donald, curious about the club situation, opened negotiations for membership in the Larchmont

Women's Club, which his grandmother had founded. He was refused, with shocked dismay. He, and some fellow construction workers, considered applying together. If refused, they planned to break down doors and smash furniture until arrested; then make a test case in court.

Unfortunately, they didn't do it, so women's right to no-men clubs may still not have been clarified. However, the question is abstract, as most men (including Donald) want nothing to do with them.

Single-sex colleges have mostly become co-ed (two-sex). Formerly all-male colleges seem to have no difficulty attracting a fifty per cent registration of females, but formerly all-female institutions may have serious problems getting enough males for their quotas.

It is a curious situation that women (or at least a large, outspoken and active group of them) are constantly pushing to join men everywhere, from playgrounds to mine labor gangs, without any marked tendency the other way. And there seems to be no Feminist thought that a man might be happier with the woman at home (and as a result, she with him) if he didn't have to be with women everywhere else. But happiness is probably unimportant where a principle is involved.

...pushing to join men everywhere...

I keep wandering away from the mice, but have kept inside the basic question of the wisdom of enforced mixing of the sexes. This question is crucial for the young, whose lives we may be warping.

Long-enduring, rewarding relationships between men and women are probably essential to human happiness. We have high and increasing rates of isolation, separation, divorce, and even mate murder. There are many causes — civilization itself is suspect — but is it right to go ahead with increasing the forced association of the sexes, when signs are plain that it may be important in our sickness?

10. Roads and Trucks

This section reports a major breakdown of the democratic process; an Act of Congress that is entirely against the country's interests, with little net advantage to even its proponents. Advance explanation is needed.

Most new highways are carefully designed for the traffic that they are intended to carry — number of vehicles per hour, and their weight, speed, and size. Number and speed refer to all vehicles, but weight and size concern trucks and buses, as automobiles are comparatively light and small.

Grades (steepness of climb and descent) are also calculated for trucks rather than cars.

Weight has been the critical factor with trucks. They are licensed for specific maximum weights, and may be very heavy. Simply supporting them when stopped or moving slowly requires thick, strong pavements and sturdy bridges. When they move at high speed they add an impact factor, particularly on rough pavements, that is really shattering. In spite of the 55 mile an hour speed limit, both cars and trucks travel at 60 to 75 in many states.

In general, the trucking industry wants to run very heavy

trucks, for more payload per mile, per trip, and particularly per driver hour. Builders and maintainers of roads want to limit truck weights severely, to reduce costly strength and maintenance requirements.

Highway systems are not just built. They are designed with care, to handle certain volumes of traffic and weight and size of vehicles.

Decisions on volume of traffic are shown chiefly in number of lanes, with exit and entrance provisions.

Lane width has been standardized at 12 feet, but is occasionally increased to 13 or 14 feet. Twelve feet is about twice the width of a car, and one and a half times the width of a truck. It has been established that vehicles need this much side clearance for safety and efficiency, both in high speed driving and maneuvering in close traffic.

Many parkways and some express roads were built before the last war with 10-foot lane widths. Some of these have been widened at great expense (and destruction of beautiful stone bridges) to 12-foot lanes, others are merely down-rated and less used.

Plans on maximum weight of the trucks to be carried are indicated to the professional eye by thickness and type of pavement, and size, strength, and spacing of beams and other supporting members in bridges carrying the through roadway and its ramps.

Design decisions become irrevocable when the high-way and its structures are built. Adding extra lanes not originally provided for usually involves total removal of overpass (highway-crossing) spans, along with retaining walls and many ramp structures, then placing longer over-passes. Highway pavements and bridges over roads or streams can usually be widened with only partial loss of existing structures.

Increasing lane width might require such rebuilding also, or space might be obtained by regrettable narrowing of safety side strips.

Adding extra lanes is usually five to ten times more costly than including them (or space for them) in the original construction, and their cost in traffic delays is enormous.

The factor of truck weight is critical, and of its size only slightly less so. In our Interstate system, maximum load figures were originally based on truck-sin-gle-trailer combinations, eight feet wide and up to 30 or 40 feet long, with maximum total weight 73,280 pounds (about 36½ tons) and maximum weight on one axle 32,000 pounds (16 tons).

... eight feet wide...

These standards were made up after considering repre-sentations from trucking people, state highway engineers, legislators, and other concerned people. They replaced individual state regulations.

It seemed that the truckers were given a fair deal on weights. And that was just the tip of the iceberg. They were also being provided with a magnificient highway system at public expense. This provided them with an increased advantage over the railroads, who must lay their own tracks and usually pay taxes on right of way.

Truckers' tax burden in fuel markup and registration fees is substantial, but probably not in proportion to highway costs, and certainly cheap for use of such a system. It is said that heavy trucks pay only half of what they should.

When weight limitations and vehicle characteristics had

been settled, design engineers could go to work. The finished roadbed (dirt or dirt-and-rock fill), in combination with the pavement that it supported, should be able to carry trucks and truck trailers at speeds up to 60 miles per hour without damage, except slight surface wear, and occasional edge breakage, for perhaps 20 years, and much longer with occasional rebuilding. Federal standards were flexible, so that in different states, concrete pavement thickness might be 8, 9, or 10 inches.

Bridges, which in ordinary sizes are usually sets of steel I-beams carrying concrete pavement directly or almost directly on their upper surfaces, were given strength for the calculated weights.

Bridge figuring was done very, very exactly, at least to four decimal places, and maybe to ten. Then a factor of safety, often 100 per cent of estimated loading, was added. This meant that the bridge was built to carry twice the estimated load, to provide for overloads plus a safety factor. It is important to keep this margin of safety, as a bridge that has been bent or otherwise damaged often needs complete replacement.

The Interstates involve so many bridges that they may have been worked out to same-spec formulae: for a four-lane 80-foot span, open Drawer D. But everything is based on the maximum weights allowed or expected, and the need for 100 per cent safety margin.

All of this was roughly agreed around 1956, when the Interstate System obtained full approval. Construction may have started then.

In 1974 Congress passed a law that permitted states to increase allowable axle loads from 16 to 17 tons, and gross weight to 40 tons. A modest increase, but why?

There are endless agruments about the extra wear and damage that trucks cause to highways, and whether trucks are paying for it. But anyhow, we had standardized truck weights, and a very expensive highway system was designed for those weights; no more. Why change?

The trucking industry seems to have sold itself on a principle that heavier trucks make more money. I look back at my time as an excavating contractor, and express doubts. In general, a moderate increase in load capacity calls for a very large increase in purchase costs of truck and tires, or a punishing increase in maintenance. When these figures are put into the books (or the computer), increased revenue is often smaller than the cost increase.

But anyhow, truckers do want more legal weight, and are prepared to fight for it, continuously.

The Interstate system is not in good condition. Skimpy maintenance, some original errors in design (mostly on the too-light, too-weak, too-narrow side) and spots of faulty construction, are showing up everywhere. This situation was noticeable in 1974; in 1982 it *... they simply do not fit.* had assumed the proportions of a disaster. By now, it looks as if thorough repair will cost more than the 60 or 70 billion for the original system, plus traffic tie-up costs.

Nevertheless, in 1982 Congress passed a bill requiring that states adopt the 1974 weight increases, which until then had been optional. It overrode all state laws to permit double trailers to operate almost everywhere, not only on the Interstates, but in thousands of places where they simply do not fit.

And, incredibly, it authorized an increase in truck width from its long-term standard of eight feet, to eight and one half feet (96 inches to 102 inches).

The first two items are clearly against the public interest, the second more than the first. But it is the third that makes our entire highway and road system obsolete — Interstate, Federal, State, and Township, plus factory loading docks and truck repair depots.

There is no immediate effect; it will appear as present trucks are replaced. For a while, truckers may be afraid to buy the wider ones, in case there is a return to sanity. But unless this permission is repealed, promptly, we will have wider trucks. This is almost the same as making our present roads narrower. And, of course, greater expense for wider lanes in roads to come.

Why should our Congress and our President do this to us? They have added weight to speed the destruction of crumbling highways, double trailers to add to traffic problems that are already a drag on the economy, and width to trucks that already seem to overfill their lanes.

There is of course the lobbying strength of the wealthy trucking industry and a powerful drivers' union. Trucking charges are a financial problem that touches most of us, and we would like to see them lower, or at least no higher. But mostly, they know not what they do.

I wish I could get these legislators to stand on some ordinary Interstate bridge over an intersecting road, and feel it rock and sway as the trucks go by. Or show them a series of crumbling pavements, and discuss the job of rebuilding them.

Because the basic problem must be that they do not understand the seriousness of the situation.

For the short term, an even more severe problem is truck overloading. A truck may be able to carry (at a price) twice its rated load, if the material is heavy and compact — gold bars, for example. Or the overweight cargo may be a single piece of a huge machine, that cannot efficiently be made smaller and lighter, or taken apart. Transportation of big items is part of proper use of the highway system. A state should check the strength of bridges and the clearance in underpasses along the route, and issue an overweight permit, accordingly.

...job of rebuilding...

This permit should be specific and limited — the load, the route, and slow speed to protect structures and the public. Too often, it is so phrased that a copy of it in each of the company's haulers protects it against weight regulations everywhere in that state. And a truck with a double load will do at least ten times the damage of a proper one.

Most deliberately overloaded trucks are not that extreme, and they are just chance-takers. Fixed weigh stations along major highways in the Northeast are preceded by a series of signs telling truckers that they MUST turn in to be weighed. But the last one almost always has a CLOSED notice hung on it.

There are mobile weigh stations that may be set up unexpectedly here and there. Truckers usually have CB radios with which they warn each other, so those with bad consciences can take detours. If you see a stream of big trucks going along some little back road, possibly over bridges marked "No Load Over Five Tons", that is why.

… big trucks…
little back road…

A driver in a familiar truck should know if it is overloaded, and might be able to make a shrewd estimate of the number of pounds. In a strange one he might, or might not. But even if knowing or suspecting, job pressures usually compel him to make the trip.

It is possible that a trucker may be tricked or pressured into taking an overload.

Overweight is a very serious problem, which is treated much too lightly except for an occasional spurt of enforcement. One of these was brought on last year by the fall of a section of the Mianus Bridge in Connecticut. That disaster had nothing to do with truck weight, but for months after it the area seemed swamped with state troopers and scales.

I am told that overloading is often encouraged by light fines for those who are caught. Truckers can average in penalties as a business expense, and load 'em up higher next trip. Enforcement crews are discouraged when fines are only a fraction of the cost of the arrests.

Judges should be aware — should be made aware — of the frequent deliberateness of violations, the scale of the money often involved, and the destruction of highways represented by the occasional violators who are caught and brought before them.

Legislators should provide bigger budgets for enforcement, and sterner measures, up to impoundment of truck and cargo, and perhaps even license suspension for the driver — but only where there is gross overloading and evidence that it is not accidental.

Sudden increase in efficiency and sternness of enforcement of regulations may carry an element of injustice. When everyone (almost) breaks a law, how can the Law select the first victims for the new regime? The best, fairest, and (fortunately) the most usual way is to start with full punishment to the worst offenders, while using lighter fines and warnings for the lesser ones.

The line between greater and lesser may then be moved downward, as enforcement becomes widespread.

Whether just or unjust, regular penalties must be severe. The factor of deterrence — reduced violations because of fear of punishment — is loud and clear. The trucking industry is closely knit, and news spreads through it readily. Enforcement results in immediate reductions of illegal loads. This operates both through owners trimming weights to keep out of trouble, and through strengthening the position of drivers reluctant to drive with overloads.

This problem must be worked out. If our highways had a voice, they would be crying for relief. Let us give it to them, quickly.

11. Odds and Ends

Nuclear Waste
Meteors
New York Bypass

Dictionaries
No Ink

Nuclear Waste

Present nuclear power plants obtain energy primarily from controlled splitting of atoms of uranium's isotope 235. The process creates a considerable bulk of highly radioactive waste, in both solid and liquid forms. Argument about precise disposal methods has prevented much action, and the material, both from power plants and weapons, is accumulating inconveniently and, in some areas, dangerously. It is expected to remain dangerous for thousands of years.

The present plan for disposal is to treat the wastes chemically when necessary, and then combine them with glass or ceramic material, that is expected to keep them solid and chemically inert, so that they will not easily dissolve in water or other earth fluids. These blocks are to be buried deeply in formations where ground water is lacking or inactive. Salt is favored for its burial caves, as it ordinarily is free of ground water, is chemically stable, and comparatively easy to dig.

However, the plan is attacked by many on varying grounds. Questions are raised about the time that glass or ceramic can contain the waste, possible invasion of formations by ground water, dissolving and carrying nuclear material, dangerously.

... *extent of the dangers is unknown.*

If my memory serves me, this problem was solved long ago, soon after it arose. It was proposed that the radioactive cubes, with some metal reinforcing and shielding, be stored on the flat floor of dry lake beds in the West. Each cube would be equipped with hoist connections, and rest on supports to hold it above the ground. The area would be criss-crossed by roads and/or railroad tracks, to provide quick access to all units by equipment.

While our present climate persists, it would be unusual for water to reach them, and such water should evaporate harmlessly, without going anywhere. If problems of disintegration or radioactive air pollution arose, the material could be easily reached for repair, shielding or removal.

It is probable that the states that include the dry lakes did not like this idea, and were able to kill it. But there seemed to be little public debate — the newspapers and scientific magazines seemed to just forget about the value of open air and access, and took to arguing about the much more severe problems of burial, and of ocean dumping.

To repeat, radioactive waste is extremely dangerous, will be for thousands of years, and full dangers are unknown.

Ocean dumping should not be even thought of, ever. Burial should be postponed until there have been many years, perhaps hundreds, of observation in the open air to fully establish the hazards.

Meteors

Meteorites are pieces of rock and/or metal that fall to the earth from space. Meteors are generally smaller pieces of similar material that are completely burned or vaporized in the atmosphere, producing visible light at night.

There are indications that many or most meteorites originate in the asteroid belt, in space between Mars and Jupiter, where the rule of planetary arrangement (Bode's Law) says that there should be a planet.

For a long time it was assumed that there had actually been a rocky, iron-cored planet in that orbit, and that it had been broken up by conflicting gravitational pulls from Jupiter, Mars, and the sun. The asteroids, and therefore our meteorites, would thus be pieces of the former planet.

Iron meteorites, often nearly pure iron and sometimes weighing many tons, seemed to prove the former existence of a planet. It is probable that masses of iron of this type can form only near the center of a substantial planet, one big enough to heat up inside.

Recently, for no clear reason (or at least, not clear to me), Scientific Opinion decided that there had been no planet, and asteroids had always been just chunks, probably starting smaller and building up. The biggest is now about 250 miles (350 km), long way.

Junking the planet...

Junking the planet brought a problem — the existence of hundreds or probably thousands of iron meteorites that must have formed in the core of the planet that they were erasing from the blackboard, and which now had no place to have come from.

A strange thing happened. Suddenly, there were no iron meteorites in fresh print. Scientific articles continued to discuss, and report findings of, stone and carbonaceous (stone-carbon mixtures) specimens, but for more than a year, the only mention of iron was when a museum moved some. Otherwise, even when completeness and accuracy demanded mention, iron was ignored, completely.

This dogma seems to be surmounting a fresh problem. It has been found that surfaces of Mars, Mercury, and our Moon show a moderately peaceful existence after formation until about four billion years ago.

Then they seemed to sustain a comparatively short, fearful battering from falling objects, whose sizes ranged from grains of sand to billions of tons.

This might well have come from breakup of a nearby planet at that time. But one recent book avoids comment on origin, another talks only about cometary orbits. Why?

There seems to be no reasonable explanation. It is as if the scientific community were an organism, capable of its own dim thought and dogma-defending reactions. Or is it just editors being tactful?

New York Bypass

Elsewhere, I have referred glowingly to the excellence of our Interstate Highway System. It is magnificient, but it has at least one local oddity — there is no freeway, nor even full-use State highway, to permit trucks from (or to) the South and Southwest to go around New York City, without substantial penalties in extra miles. Their best crossing of the Hudson River is the George Washington Bridge, with miles of heavily congested in-city freeways to the east.

In general, the System's layout has wisely included provision for traffic flow around big cities. But not New York.

On the map, it looks possible, and even easy, to provide such a drive-around by continuing the New Jersey Turnpike (I-95) north of its present end at I-80 (Washington Bridge) to the New York Thruway (I-87), a distance of about fifteen miles, in two states. The Thruway, and its nearby Tappan Zee Bridge across the Hudson, are outside the city.

The Turnpike could have been a disaster for parts of the area it would have crossed, but scars would have been

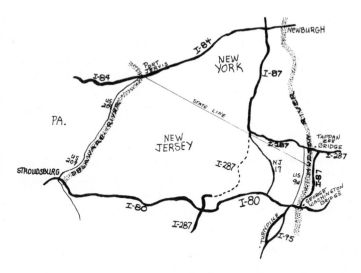

mostly healed by now. Interstate history shows other pretty villages that were casually mangled when they were in the way, sometimes for less important reasons.

An attempt was made, a little too far west for general convenience, to link South and West with the Tappan Zee by means of I-287. But the conservationists opposed it on some ground that I cannot remember, and for ten or twenty years an 18-mile section of it has been only a double-dash line on New Jersey maps. From far-off Connecticut, I assumed that a compromise could and would be reached, but that road is still unbuilt.

There is US 9W, that runs directly north from the Washington Bridge to a ramp to the Tappan Zee. But trucks over 10 tons are prohibited in its New York section, probably because of a high and slender viaduct in Piermont, and general under-design for heavy traffic.

For a while, trucks were using Pennsylvania 209, alongside the Delaware River, between I-80 (to the Washington Bridge) and I-84, which has its own crossing at Newburgh, north of the Tappan Zee. But 209 is narrow for heavy trucks, so the National Park Service banned them.

Highway 17 connects these two Interstates between Hackensack and Suffern, and is a heavy duty road. But it has a strong slant, northwest to southeast, and traffic problems of its own. Transfer would cost an eastbound truck on route 80, or a westbound one on 84, about twelve extra miles. Most drivers prefer to take their chance with congestion east of the Washington Bridge.

The Turnpike extension seems so necessary that it is hard to believe that it wasn't built; when that highway had its last overhaul. An arrangement to finish 287 would certainly help. In the meantime, this uneconomic and potentially dangerous situation should get more public attention.

Dictionaries

For most widely used languages, there are translating dictionaries. In them, the principal words of each of two languages are listed alphabetically in separate sections. Each word is defined and explained in the other language. There may or may not also be pages of explanations of irregularities and grammar in one or both tongues.

Why are all these words omitted?

A person unfamiliar with a language should be able to work out the meaning of a simple instruction, or a newspaper paragraph, by looking its words up one after another and stringing them together. Such a translation would be rough, but it might be expected to be usable.

Unfortunately, publishers of most of these dictionaries have a policy of including only the infinitive form of a verb. If the verb is irregular, the form to be translated may have no resemblance whatever to its infinitive form. The principal offender in a whole group of languages is 'be', which not only has a wide variety of apparently unrelated forms, but also appears very frequently in most written material. In English these include is, was, were, am, are.

Lack of understanding of just this one word in its various forms often makes translation impossible.

Some of the larger dictionaries give these forms in a special irregular verb section. But even if our would-be translator has one of these, he may not know it or may not think of it. He will look for words only where they belong — in alphabetical listing — and assume from their lack that the dictionary is incomplete and therefore defective.

Why are all these words omitted? I have had it explained to me, and still wish I knew. I have warm personal relations with a Mexican publisher who produces dictionaries. I came to them prattling happily of my wonderful idea of including all the irregular verb forms (there are only a few dozen) to produce a true translating dictionary that would be more useful and therefore sell better.

I was shushed and diverted from the subject in a shocked but forgiving pity-the-poor-foreigner manner. The atmosphere was as strained as if I had started to talk about sex at a dinner table in the year 1900.

An explanation was offered, but it seemed, embarrassed, complicated, unclear, and unimportant.

Snub or no snub, it is still my conviction that every translating dictionary should include every common word and word form, regardless of any grief that they might cause to grammarians. People using it are more important.

No Ink

Many of our tasks have been taken over by computer, and their cousins, calculators and cash registers. The last two are my interest here.

In general, they work well, and cost far less than the motor or even hand crank types that preceded them.

Calculators are much more versatile than the adding machine was. They cheerfully multiply, divide, juggle decimal point position, and take percentages. For a few dollars extra, they may do square root, or help Junior with his trigonometry. Some even have memories.

But they have their drawbacks. Instead of separate CLEAR keys for the wrong number you just struck and the number you just added, they have put them on one key, CE/E, and either it or I often get confused. Also, any time I touch ANY key, the paper tape is likely to jump two or three spaces, to help build tangles high on the desk.

... black is not sporting.

There are individual tricks too. Mine was so enthusiastic about multiplying that it would sometimes multiply the product by two. I saved a couple of the tapes on which it did (they didn't show the 2, just the doubling). But like most of my papers, they disappeared. And the machine reformed — I think.

But the point I have been trying to get to is printing quality. My office is well lit, so with my glasses I can read the tape fine. But if I want to take a quick look at it in the car, I can just forget it. In the old days there was a period, at least a couple of months and sometimes a year, when the adding machine printed as black (or red) as a typewriter. And when I took the ribbon out for replacement, it was still printing darker than a new one can now.

The economics of the calculator business has never been explained to me. I am just guessing. But if the old ones somehow found that it paid to have two clear keys, with CE on one and C on the other, with a readable tape, for $200, couldn't my electronic for $40 keep those assets, for perhaps $5 more, and still be a bargain?

Sometimes there is a big problem with the printing device, because it doesn't have a ribbon. I bought a beautiful little electronic portable typewriter that weighed only five pounds. Its problem was that it couldn't make carbon copies, and its first copies were so faint I could hardly read them. My copy people ducked out the back door when I brought them there.

This electronic typewriter's printing device sprayed blue dots, very faint, to build its letters. It might have done much better with black, but in this field, if there is a problem of print intensity, black ink is not sporting.

Lots of computers use spray printers, blue, or course. Their makers say that such printers are cheap and fast, and hardly anyone reads the garbage that they print anyhow.

But I must keep away from computers, and get to their near (and getting nearer) relatives, cash registers.

I don't like them much, but the stores change models so fast that it is hard to work up a good grudge. I meet them mostly in supermarket groceries, where it seems important in customer relations to have the latest and the best.

The better (or at least most friendly) registers have a side display where the customer can see each amount as it is rung, and also the item in the checker's hand. If she (oops, a sex assumption) is sharp (much sharper than I am), she can compare the amount with the price posted in yesterday's ad, and/or the price on the package, and make a protest if it is necessary.

I'm not with it on item prices. What I want and need, and never get, is a running total, showing what I owe, below the displayed price. My calculations are concerned with whether my last twenty is enough, or should I dig out a check, along with a pen and identification?

Many quite new registers do not have the side display, or the store piles merchandise to hide it. If the customer wants to watch, she (sorry, I can't stop it) can go to the end of the counter, and watch over the checker's shoulder.

The most sophisticated register (so far) reads a line pattern on the package, tells the checker and the customer in a clear voice the description of the item and its price, and prints the same information on a tape.

All of these registers, from the simplest to the I-can't-quite-believe-it talker, print a paper tape showing a least the price of each item and the total. But — and this is the point of this article — any tape, from the highest to the lowest machine — is more than likely to be printed so faintly that reading it is difficult.

I am told that the price of cash registers range from $600 to over $10,000, and that ribbons cost from three to four dollars each.